SMART GRID

SMART GRID
TECHNOLOGY AND APPLICATIONS

Janaka Ekanayake
Cardiff University, UK

Kithsiri Liyanage
University of Peradeniya, Sri Lanka

Jianzhong Wu
Cardiff University, UK

Akihiko Yokoyama
University of Tokyo, Japan

Nick Jenkins
Cardiff University, UK

WILEY

A John Wiley & Sons, Ltd., Publication

Library of Congress Cataloging-in-Publication Data

Smart grid : technology and applications / Janaka Ekanayake . . . [et al.].
    p. cm.
  Includes bibliographical references and index.
  ISBN 978-0-470-97409-4 (cloth)
 1. Smart power grids.   I. Ekanayake, J. B. (Janaka B.)
  TK3105.S677 2012
  621.31–dc23

                                    2011044006

A catalogue record for this book is available from the British Library.

Print ISBN: 978-0-470-97409-4

Typeset in 10/12pt Times by Aptara Inc., New Delhi, India.

Contents

Part III   POWER ELECTRONICS AND ENERGY STORAGE

About the Authors

Janaka Ekanayake received his BSc Eng Degree in Electrical and Electronic Engineering from the University of Peradeniya, Sri Lanka, in 1990 and his PhD in Electrical Engineering from the University of Manchester Institute of Science and Technology (UMIST), UK in 1995. He is presently a Senior Lecturer at Cardiff University, UK. Prior to that he was a Professor in the Department of Electrical and Electronic Engineering, University of Peradeniya. His main research interests include power electronic applications for power systems, renewable energy generation and its integration. He is a Chartered Engineer, a Fellow of the IET, a Senior Member of IEEE, and a member of the IESL. He has published more than 30 papers in refereed journals and has also co-authored three books.

Kithsri M. Liyanage is attached to the Department of Electrical and Electronic Engineering, University of Peradeniya, Sri Lanka, as a Professor. He obtained his BSc Eng from the University of Peradeniya in 1983 and his Dr Eng from the University of Tokyo in 1991. He was a Visiting Scientist at the Department of Electrical Engineering, the University of Washington, from 1993 to 1994 and a Visiting Research Fellow at the Advanced Centre for Power and Environmental Technology, the University of Tokyo, Japan, from 2008 to 2010. He has authored or co-authored more than 30 papers related to Smart Grid applications and control since 2009. His research interest is mainly in the application of ICT for the realisation of the Smart Grid.

Jianzhong Wu received his BSc, MSc and PhD in 1999, 2001 and 2004 respectively, from Tianjin University, China. He was an Associate Professor in Tianjin University, and then moved to the University of Manchester as a research fellow in 2006. Since 2008, he has been a lecturer at the Cardiff School of Engineering. His main research interests include Energy Infrastructure and Smart Grids. He has a track record of undertaking a number of EU and other funded projects. He is a member of the IET, the IEEE and the ACM. He has published more than 30 papers and co-authored one book.

Akihiko Yokoyama received his BS, MS and PhD in 1979, 1981 and 1984 respectively, from the University of Tokyo, Japan. Since 2000, he has been a Professor in the Department of Electrical Engineering, the University of Tokyo. He has been a Visiting Scholar at the University of Texas at Arlington and the University of California at Berkeley. His main research interests include power system analysis and control and Smart Grids. He is a Senior

Member of the Institute of Electrical Engineers of Japan (IEEJ), the Japan Society for Industrial and Applied Mathematics (JSIAM), the IEEE and a member of CIGRE.

Nick Jenkins was at the University of Manchester (UMIST) from 1992 to 2008. He then moved to Cardiff University where he is now Professor of Renewable Energy. His previous career had included 14 years industrial experience, of which five years were in developing countries. While at Cardiff University he has developed teaching and research activities in electrical power engineering and renewable energy. He is a Fellow of the IET, the IEEE and the Royal Academy of Engineering. He is a Distinguished Member of CIGRE and from 2009 to 2011 was the Shimizu Visiting Professor to the Atmosphere and Energy Program at Stanford University, USA.

Preface

Electric power systems throughout the world are facing radical change stimulated by the pressing need to decarbonise electricity supply, to replace ageing assets and to make effective use of rapidly developing information and communication technologies (ICTs). These aims all converge in the Smart Grid. The Smart Grid uses advanced information and communication to control this new energy system reliably and efficiently. Some ICT infrastructure already exists for transmission voltages but at present there is very little real-time communication either to or from the customer or in distribution circuits.

The Smart Grid vision is to give much greater visibility to lower voltage networks and to enable the participation of customers in the operation of the power system, particularly through Smart Meters and Smart Homes. The Smart Grid will support improved energy efficiency and allow a much greater utilisation of renewables. Smart Grid research and development is currently well funded in the USA, the UK, China, Japan and the EU. It is an important research topic in all parts of the world and the source of considerable commercial interest.

The aim of the book is to provide a basic discussion of the Smart Grid concept and then, in some detail, to describe the technologies that are required for its realisation. Although the Smart Grid concept is not yet fully defined, the book will be valuable in describing the key enabling technologies and thus permitting the reader to engage with the immediate development of the power system and take part in the debate over the future of the Smart Grid.

This book is the outcome of the authors' experience in teaching to undergraduate and MSc students in China, Japan, Sri Lanka, the UK and the USA and in carrying out research. The content of the book is grouped into three main technologies:

1. Part I Information and communication systems (Chapters 2–4)
2. Part II Sensing, measurement, control and automation (Chapters 5–8)
3. Part III Power electronics and energy storage (Chapters 9–12).

These three groups of technologies are presented in three Parts in this book and are relatively independent of each other. For a course module on an MEng or MSc in power systems or energy Chapters 2-4, 5-7 and 9-11 are likely to be most relevant, whereas for a more general module on the Smart Grid, Chapters 2–5 and Chapters 9 and 12 are likely to be most appropriate.

The technical content of the book includes specialised topics that will appeal to engineers from various disciplines looking to enhance their knowledge of technologies that are making an increasing contribution to the realisation of the Smart Grid.

Acknowledgements

We would like to acknowledge contributions from colleagues and individuals without whom this project will not be a success. Particular thanks are due to Toshiba, S&C Electric Europe Ltd., Tokyo Electric Power Co., Japan Wind and Tianda Qiushi Power New Technology Co. Ltd. for generously making available a number of photographs. Also we would like to thank John Lacari for checking the numerical examples; Jun Liang, Lee Thomas, Alasdair Burchill, Panagiotis Papadopoulos, Carlos Ugalde-Loo and Iñaki Grau for providing information for Chapters 5, 11 and 12; Mahesh Sooriyabandara for checking some chapters; and Luke Livermore, Kamal Samarakoon, Yan He, Sugath Jayasinghe and Bieshoy Awad who helped in numerous ways.

List of Abbreviations

2-D	2-dimensional
3-D	3-dimensional
3G	3rd Generation mobile systems
3GPP	3rd Generation Partnership Project
ACL	Asynchronous Connectionless Link
ADC	Analogue to Digital Conversion or Converter
ADMD	After Diversity Maximum Demand
ADSL	Asymmetric Digital Subscriber Line
ADSS	All-Dielectric Self-Supporting
AES	Advanced Encryption Standard
AGC	Automatic Generation Control
AM	Automated Mapping
AMM	Automatic Meter Management
AMR	Automatic Meter Reading
ARIMA	Autoregressive Integrated Moving Average
ARIMAX	Autoregressive Integrated Moving Average with exogenous variables
ARMA	Autoregressive Moving Average
ARMAX	Autoregressive Moving Average with exogenous variables
ARPANET	Advanced Research Projects Agency Network
ASDs	Adjustable Speed Drives
ASK	Amplitude Shift Keying
AVC	Automatic Voltage Control
BES	Battery Energy Storage
BEV	Battery Electric Vehicles
BPL	Broadband over Power Line
CB	Circuit Breaker
CC	Constant Current
CI	Customer Interruptions
CIM	Common Information Model
CIS	Customer Information System
CML	Customer Minutes Lost
COSEM	Companion Specification for Energy Metering
CSC	Current Source Converter
CSC-HVDC	Current Source Converter High Voltage DC

CSMA/CD	Carrier Sense Multiple Access/Collision Detect
CT	Current Transformer
CTI	Computer Telephony Integration
CV	Constant Voltage
CVT	Capacitor Voltage Transformers
DAC	Digital to Analogue Converter
DARPA	Defense Advanced Research Project Agency
DB	Demand Bidding
DCC	Diode-Clamped Converter
DER	Distributed Energy Resources
DES	Data Encryption Standard
DFIG	Doubly Fed Induction Generators
DG	Distributed Generation
DLC	Direct Load Control
DMS	Distribution Management System
DMSC	Distribution Management System Controller
DNO	Distribution Network Operators
DNS	Domain Name Server
DR	Demand Response
DSB	Demand-Side Bidding
DSI	Demand-Side Integration
DSL	Digital Subscriber Lines
DSM	Demand-Side Management
DSP	Digital Signal Processor
DSR	Demand-Side Response
DVR	Dynamic Voltage Restorer
EDGE	Enhanced Data Rates for GSM Evolution
EMI	Electromagnetic Interference
EMS	Energy Management System
ESS	Extended Service Set
EU	European Union
EV	Electric Vehicles
FACTS	Flexible AC Transmission Systems
FCL	Fault Current Limiters
FCS	Frame Check Sequence
FFD	Full Function Device
FM	Facilities Management
FPC	Full Power Converter
FSIG	Fixed Speed Induction Generator
FSK	Frequency Shift Keying
FTP	File Transfer Protocol
GEO	Geostationary Orbit
GGSN	Gateway GPRS Support Node
GIS	Gas Insulated Substations
GIS	Geographic Information System
GPRS	General Packet Radio Service

GPS	Global Positioning System
GSM	Global System for Mobile Communications
GTO	Gate Turn-off (Thyristor)
HAN	Home-Area Network
HDLC	High-Level Data Link Control
HMI	Human Machine Interface
HTTP	Hypertext Transfer Protocol
HVAC	Heating, Ventilation, Air Conditioning
HVDC	High Voltage DC
ICT	Information and Communication Technology
IED	Intelligent Electronic Device
IGBT	Insulated Gate Bipolar Transistor
IGCT	Insulated Gate Commutated Thyristor
IP	Internet Protocol
IPFC	Interline Power Flow Controller
IPng	IP Next Generation
IPsec	Internet Protocol Security
ITE	Information Technology Equipment
KDC	Key Distribution Centre
LAN	Local Area Network
LCD	Liquid Crystal Displays
LED	Light Emitting Diodes
LLC	Logical Link Control
LMU	Line Matching Unit
LOLP	Loss of Load Probability
M2C	Multi-Modular Converter
MAS	Multi Agent System
MD	Message Digest
MDM	Metre Data Management system
METI	Ministry of Economy, Trade and Industry
MGCC	MicroGrid Central Controllers
MMS	Manufacturing Message Specification
MOSFET	Metal Oxide Semiconductor Field Effect Transistor
MPLS	Multi Protocol Label Switching
MPPT	Maximum Power Point Tracking
MSB	Most Significant Bit
MTSO	Mobile Telephone Switching Office
NAN	Neighbourhood Area Network
NERC CIP	North America Electric Reliability Corporation – Critical Infrastructure Protection
NOP	Normally Open Point
NPC	Neutral-Point-Clamped
OCGT	Open Cycle Gas Turbines
OFDM	Orthogonal Frequency Multiplexing
OFDMA	Orthogonal Frequency Division Multiple Access
OLTCs	On-Load Tap Changers

OMS	Outage Management System
OPGW	OPtical Ground Wires
PCM	Pulse Code Modulation
PDC	Phasor Data Concentrator
PET	Polyethylene Terephathalate
PGA	Programmable Gain Amplifier
PHEV	Plug-in Hybrid Electric Vehicles
PLC	Power Line Carrier
PLL	Phase Locked Loop
PMU	Phasor Measurement Units
PSK	Phase Shift Keying
PSS	Power System Stabilisers
PSTN	Public Switched Telephone Network
PV	Photovoltaic
PWM	Pulse Width Modulation
RFD	Reduced Function Device
RMU	Ring Main Unit
RTU	Remote Terminal Unit
SAP	Session Announcement Protocol
SCADA	Supervisory Control and Data Acquisition
SCE	Southern California Edison
SCO	Synchronous Connection Orientated
SGCC	State Grid Corporation of China
SGSN	Serving GPRS Support Node
SHA	Secure Hash Algorithm
SMES	Superconducting Magnetic Energy Storage
SMTP	Simple Mail Transfer Protocol
SNR	Signal to Noise Ratio
SOC	State Of Charge
SVC	Static Var Compensator
TCP	Transmission Control Protocol
TCR	Thyristor Controlled Reactor
TCSC	Thyristor Controlled Series Capacitor
THD	Total Harmonic Distortion
TSC	Thyristor Switched Capacitor
TSSC	Thyristor Switched Series Capacitor
UHV	Ultra High Voltage
UML	Unified Modelling Language
UPFC	Unified Power Flow Controller
UPS	Uninterruptable Power Supplies
URL	Uniform Resource Locator
UTP	Unshielded Twisted Pair
VPN	Virtual Private Network
VPP	Virtual Power Plant
VSC	Voltage Source Converter
VSC-ES	Voltage Source Converters with Energy Storage

VSC-HVDC	Voltage Source Converter HVDC
VT	Voltage Transformer
WAMPAC	Wide Area Monitoring, Protection and Control
WAMSs	Wide-Area Measurement Systems
WAN	Wide Area Network
WiMax	Worldwide Interoperability for Microwave Access
WLAN	Wireless LAN
WLAV	Weighted Least Absolute Value
WLS	Weighted Least Square
WPAN	Wireless Public Area Networks
XOR	Exclusive OR

1

The Smart Grid

1.1 Introduction

Established electric power systems, which have developed over the past 70 years, feed electrical power from large central generators up through generator transformers to a high voltage inter-connected network, known as the transmission grid. Each individual generator unit, whether powered by hydropower, nuclear power or fossil fuelled, is large with a rating of up to 1000 MW. The transmission grid is used to transport the electrical power, sometimes over consid-erable distances, and this power is then extracted and passed through a series of distribution transformers to final circuits for delivery to the end customers.

The part of the power system supplying energy (the large generating units and the transmis-sion grid) has good communication links to ensure its effective operation, to enable market transactions, to maintain the security of the system, and to facilitate the integrated operation of the generators and the transmission circuits. This part of the power system has some automatic control systems though these may be limited to local, discrete functions to ensure predictable behaviour by the generators and the transmission network during major disturbances.

The distribution system, feeding load, is very extensive but is almost entirely passive with little communication and only limited local controls. Other than for the very largest loads (for example, in a steelworks or in aluminium smelters), there is no real-time monitoring of either the voltage being offered to a load or the current being drawn by it. There is very little interaction between the loads and the power system other than the supply of load energy whenever it is demanded.

The present revolution in communication systems, particularly stimulated by the internet, offers the possibility of much greater monitoring and control throughout the power system and hence more effective, flexible and lower cost operation. The Smart Grid is an opportunity to use new ICTs (Information and Communication Technologies) to revolutionise the electrical power system. However, due to the huge size of the power system and the scale of investment that has been made in it over the years, any significant change will be expensive and requires careful justification.

The consensus among climate scientists is clear that man-made greenhouse gases are leading to dangerous climate change. Hence ways of using energy more effectively and generating electricity without the production of CO_2 must be found. The effective management of loads

Smart Grid: Technology and Applications, First Edition.
Janaka Ekanayake, Kithsiri Liyanage, Jianzhong Wu, Akihiko Yokoyama and Nick Jenkins.
© 2012 John Wiley & Sons, Ltd. Published 2012 by John Wiley & Sons, Ltd.

and reduction of losses and wasted energy needs accurate information while the use of large amounts of renewable generation requires the integration of the load in the operation of the power system in order to help balance supply and demand. Smart meters are an important element of the Smart Grid as they can provide information about the loads and hence the power flows throughout the network. Once all the parts of the power system are monitored, its state becomes observable and many possibilities for control emerge.

In the UK, the anticipated future de-carbonised electrical power system is likely to rely on generation from a combination of renewables, nuclear generators and fossil-fuelled plants with carbon capture and storage. This combination of generation is difficult to manage as it consists of variable renewable generation and large nuclear and fossil generators with carbon capture and storage that, for technical and commercial reasons, will run mainly at constant output. It is hard to see how such a power system can be operated cost-effectively without the monitoring and control provided by a Smart Grid.

1.2   Why implement the Smart Grid now?

Since about 2005, there has been increasing interest in the Smart Grid. The recognition that ICT offers significant opportunities to modernise the operation of the electrical networks has coincided with an understanding that the power sector can only be de-carbonised at a realistic cost if it is monitored and controlled effectively. In addition, a number of more detailed reasons have now coincided to stimulate interest in the Smart Grid.

1.2.1   Ageing assets and lack of circuit capacity

In many parts of the world (for example, the USA and most countries in Europe), the power system expanded rapidly from the 1950s and the transmission and distribution equipment that was installed then is now beyond its design life and in need of replacement. The capital costs of like-for-like replacement will be very high and it is even questionable if the required power equipment manufacturing capacity and the skilled staff are now available. The need to refurbish the transmission and distribution circuits is an obvious opportunity to innovate with new designs and operating practices.

In many countries the overhead line circuits, needed to meet load growth or to connect renewable generation, have been delayed for up to 10 years due to difficulties in obtaining rights-of-way and environmental permits. Therefore some of the existing power transmission and distribution lines are operating near their capacity and some renewable generation cannot be connected. This calls for more intelligent methods of increasing the power transfer capacity of circuits dynamically and rerouting the power flows through less loaded circuits.

1.2.2   Thermal constraints

Thermal constraints in existing transmission and distribution lines and equipment are the ultimate limit of their power transfer capability. When power equipment carries current in excess of its thermal rating, it becomes over-heated and its insulation deteriorates rapidly. This leads to a reduction in the life of the equipment and an increasing incidence of faults.

If an overhead line passes too much current, the conductor lengthens, the sag of the catenary increases, and the clearance to the ground is reduced. Any reduction in the clearance of an overhead line to the ground has important consequences both for an increase in the number of faults but also as a danger to public safety. Thermal constraints depend on environmental conditions, that change through the year. Hence the use of dynamic ratings can increase circuit capacity at times.

1.2.3 Operational constraints

Any power system operates within prescribed voltage and frequency limits. If the voltage exceeds its upper limit, the insulation of components of the power system and consumer equipment may be damaged, leading to short-circuit faults. Too low a voltage may cause malfunctions of customer equipment and lead to excess current and tripping of some lines and generators. The capacity of many traditional distribution circuits is limited by the variations in voltage that occur between times of maximum and minimum load and so the circuits are not loaded near to their thermal limits. Although reduced loading of the circuits leads to low losses, it requires greater capital investment.

Since about 1990, there has been a revival of interest in connecting generation to the distribution network. This distributed generation can cause over-voltages at times of light load, thus requiring the coordinated operation of the local generation, on-load tap changers and other equipment used to control voltage in distribution circuits. The frequency of the power system is governed by the second-by-second balance of generation and demand. Any imbalance is reflected as a deviation in the frequency from 50 or 60 Hz or excessive flows in the tie lines between the control regions of very large power systems. System operators maintain the frequency within strict limits and when it varies, response and reserve services are called upon to bring the frequency back within its operating limits [1]. Under emergency conditions some loads are disconnected to maintain the stability of the system.

Renewable energy generation (for example. wind power, solar PV power) has a varying output which cannot be predicted with certainty hours ahead. A large central fossil-fuelled generator may require 6 hours to start up from cold. Some generators on the system (for example, a large nuclear plant) may operate at a constant output for either technical or commercial reasons. Thus maintaining the supply–demand balance and the system frequency within limits becomes difficult. Part-loaded generation 'spinning reserve' or energy storage can address this problem but with a consequent increase in cost. Therefore, power system operators increasingly are seeking frequency response and reserve services from the load demand. It is thought that in future the electrification of domestic heating loads (to reduce emissions of CO_2) and electric vehicle charging will lead to a greater capacity of flexible loads. This would help maintain network stability, reduce the requirement for reserve power from part-loaded generators and the need for network reinforcement.

1.2.4 Security of supply

Modern society requires an increasingly reliable electricity supply as more and more critical loads are connected. The traditional approach to improving reliability was to install additional redundant circuits, at considerable capital cost and environmental impact. Other than disconnecting the faulty circuit, no action was required to maintain supply after a fault. A Smart Grid approach is to use intelligent post-fault reconfiguration so that after the (inevitable) faults in the power system, the supplies to customers are maintained but to avoid the expense of multiple circuits that may be only partly loaded for much of their lives. Fewer redundant circuits result in better utilisation of assets but higher electrical losses.

1.2.5   National initiatives

Many national governments are encouraging Smart Grid initiatives as a cost-effective way to modernise their power system infrastructure while enabling the integration of low-carbon energy resources. Development of the Smart Grid is also seen in many countries as an important economic/commercial opportunity to develop new products and services.

1.2.5.1   China

The Chinese government has declared that by 2020 the carbon emission per-unit of GDP will reduce to 40~45 per cent of that in 2008. Other drivers for developing the Smart Grid in China are the nation's rapid economic growth and the uneven geographical distribution of electricity generation and consumption.

The State Grid Corporation of China (SGCC) has released a medium–long term plan of the development of the Smart Grid. The SGCC interprets the Smart Grid [2] as

> "*a strong and robust electric power system, which is backboned with Ultra High Voltage (UHV) networks; based on the coordinated development of power grids at different voltage levels; supported by information and communication infrastructure; characterised as an automated, and interoperable power system and the integration of electricity, information, and business flows.*"

1.2.5.2   The European Union

The SmartGrids Technology Platform of the European Union (EU) has published a vision and strategy for Europe's electricity networks of the future [3]. It states:

> "*It is vital that Europe's electricity networks are able to integrate all low carbon generation technologies as well as to encourage the demand side to play an active part in the supply chain. This must be done by upgrading and evolving the networks efficiently and economically.*"

The SmartGrids Technology Platform identified the following important areas as key challenges that impact on the delivery of the EU-mandated targets for the utilisation of renewable energy, efficiency and carbon reductions by 2020 and 2050:

- strengthening the grid, including extending it offshore;
- developing decentralised architectures for system control;
- delivering communications infrastructure;
- enabling an active demand side;
- integrating intermittent generation;

- enhancing the intelligence of generation, demand and the grid;
- capturing the benefits of distributed generation (DG) and storage;
- preparing for electric vehicles.

1.2.5.3 Japan

In 2009, the Japanese government declared that by 2020 carbon emissions from all sectors will be reduced to 75 per cent of those in 1990 or two-thirds of those in 2005. In order to achieve this target, 28 GW and 53 GW of photovoltaic (PV) generations are required to be installed in the power grid by 2020 and 2030. The Ministry of Economy, Trade and Industry (METI) has set up three study committees since 2008 to look into the Smart Grid and related aspects. These committees were active for a one-year period and were looking at the low-carbon power system (2008–2009), the next-generation transmission and distribution network, the Smart Grid in the Japanese context (2009–2010) and regulatory issues of the next-generation transmission and distribution system (2010–2011). The mandate given to these committees was to discuss the following technical and regulatory issues regarding the large penetration of renewable energy, especially PV generation, into the power grid:

- surplus power under light load conditions;
- frequency fluctuations;
- voltage rise on distribution lines;
- priority interconnection, access and dispatching for renewable energy-based generators;
- cost recovery for building the Smart Grid.

Further, a national project called 'The Field Test Project on Optimal Control Technologies for the Next-Generation Transmission and Distribution System' was conducted by 26 electric utilities, manufacturing companies and research laboratories in Japan in order to develop the technologies to solve these problems.

Since the Tohoku earthquake on 11 March 2011, the Smart Grid has been attracting much attention for the reconstruction of the damaged districts and the development of a low-carbon society.

1.2.5.4 The UK

The Department of Energy and Climate Change document *Smarter Grids: The Opportunity* [4] states that the aim of developing the Smart Grid is to provide flexibility to the current electricity network, thus enabling a cost-effective and secure transition to a low-carbon energy system. The Smart Grid route map [5] recognises a number of critical developments that will drive the UK electrical system towards a low carbon system. These include:

- rapid expansion of intermittent renewables and less flexible nuclear generation in conjunction with the retirement of flexible coal generation;
- electrification of heating and transport;
- penetration of distributed energy resources which include distributed generation, demand response and storage;
- increasing penetration of electric vehicles.

1.2.5.5 The USA

According to Public Law 110–140-DEC. 19, 2007 [6], the United States of America (the USA)

"is supporting modernisation of the electricity transmission and distribution networks to maintain a reliable and secure electricity infrastructure that can meet future demand growth and to achieve increased use of digital information and controls technology; dynamic optimisation of grid operations and resources; deployment and integration of distributed resources and generation; development and incorporation of demand response, demand-side resources, and energy-efficient resources; development of 'smart' technologies for metering, communications and status, and distribution automation; integration of 'smart' appliances and consumer devices; deployment and integration of advanced electricity storage and peak-shaving technologies; provisions to consumers of timely information and control options and development of standards for communication and inter-operability."

1.3 What is the Smart Grid?

The Smart Grid concept combines a number of technologies, end-user solutions and addresses a number of policy and regulatory drivers. It does not have a single clear definition.

The European Technology Platform [3] defines the Smart Grid as:

"A SmartGrid is an electricity network that can intelligently integrate the actions of all users connected to it – generators, consumers and those that do both – in order to efficiently deliver sustainable, economic and secure electricity supplies."

According to the US Department of Energy [7]:

"A smart grid uses digital technology to improve reliability, security, and efficiency (both economic and energy) of the electric system from large generation, through the delivery systems to electricity consumers and a growing number of distributed-generation and storage resources."

In *Smarter Grids: The Opportunity* [4], the Smart Grid is defined as:

"A smart grid uses sensing, embedded processing and digital communications to enable the electricity grid to be observable (able to be measured and visualised), controllable (able to be manipulated and optimised), automated (able to adapt and self-heal), fully integrated (fully interoperable with existing systems and with the capacity to incorporate a diverse set of energy sources)."

The literature [7–10] suggests the following attributes of the Smart Grid:

1. It enables demand response and demand side management through the integration of smart meters, smart appliances and consumer loads, micro-generation, and electricity storage (electric vehicles) and by providing customers with information related to energy use and prices. It is anticipated that customers will be provided with information and incentives to modify their consumption pattern to overcome some of the constraints in the power system.
2. It accommodates and facilitates all renewable energy sources, distributed generation, residential micro-generation, and storage options, thus reducing the environmental impact

of the whole electricity sector and also provides means of aggregation. It will provide simplified interconnection similar to 'plug-and-play'.
3. It optimises and efficiently operates assets by intelligent operation of the delivery system (rerouting power, working autonomously) and pursuing efficient asset management. This includes utilising asserts depending on what is needed and when it is needed.
4. It assures and improves reliability and the security of supply by being resilient to disturbances, attacks and natural disasters, anticipating and responding to system disturbances (predictive maintenance and self-healing), and strengthening the security of supply through enhanced transfer capabilities.
5. It maintains the power quality of the electricity supply to cater for sensitive equipment that increases with the digital economy.
6. It opens access to the markets through increased transmission paths, aggregated supply and demand response initiatives and ancillary service provisions.

1.4 Early Smart Grid initiatives

1.4.1 Active distribution networks

Figure 1.1 is a schematic of a simple distribution network with distributed generation (DG). There are many characteristics of this network that differ from a typical passive distribution network. First, the power flow is not unidirectional. The direction of power flows and the voltage magnitudes on the network depend on both the demand and the injected generation. Second, the distributed generators give rise to a wide range of fault currents and hence complex protection and coordination settings are required to protect the network. Third, the

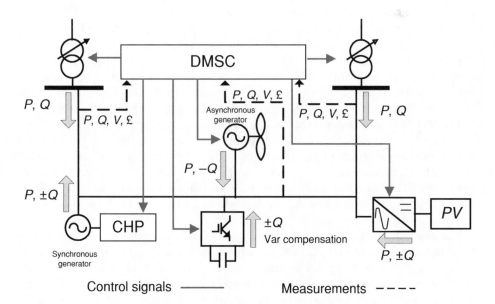

Figure 1.1 Distribution network active management scheme

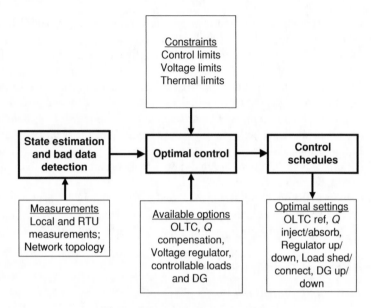

Figure 1.2   Architecture of a DMSC

reactive power flow on the network can be independent of the active power flows. Fourth, many types of DGs are interfaced through power electronics and may inject harmonics into the network.

Figure 1.1 also shows a control scheme suitable for achieving the functions of active control. In this scheme a Distribution Management System Controller (DMSC) assesses the network conditions and takes action to control the network voltages and flows. The DMSC obtains measurements from the network and sends signals to the devices under its control. Control actions may be a transformer tap operation, altering the DG output and injection/absorption of reactive power.

Figure 1.2 shows the DMSC controller building blocks that assess operating conditions and find the control settings for devices connected to the network. The key functions of the DMSC are state estimation, bad data detection and the calculation of optimal control settings.

The DMSC receives a limited number of real-time measurements at set intervals from the network nodes. The measurements are normally voltage, load injections and power flow measurements from the primary substation and other secondary substations. These measurements are used to calculate the network operating conditions. In addition to these real-time measurements, the DMSC uses load models to forecast load injections at each node on the network for a given period that coincides with the real-time measurements. The network topology and impedances are also supplied to the DMSC.

The state estimator (described in Chapter 7) uses this data to assess the network conditions in terms of node voltage magnitudes, line power flows and network injections. Bad measurements coming to the system will be filtered using bad data detection and identification methods.

When the network operating conditions have been assessed, the control algorithm identifies whether the network is operating within its permissible boundaries. This is normally assessed by analysing the network voltage magnitudes at each busbar. The optimisation algorithm is supplied with the available active control options, the limits on these controls and the network operating constraints. Limits on controls are the permissible lower and higher settings of the equipment. Operating constraints are usually voltage limits and thermal ratings of the lines and equipment. The optimal control algorithm calculates the required control settings and optimises the device settings without violating constraints and operating limits.

The solution from the control algorithm is the optimal control schedules that are sent to the devices connected to the network. Such control actions can be single or multiple control actions that would alter the set point of any of the devices by doing any of the following:

- alter the reference of an On-Load Tap Changer (OLTC) transformer/voltage regulator relay;
- request the Automatic Voltage Regulator (AVR) or the governor of a synchronous generator to alter the reactive/active power of the machine;
- send signals to a wind farm Supervisory Control and Data Acquisition (SCADA) system to decrease the wind farm output power;
- shed or connect controllable loads on the network;
- increase or decrease the settings of any reactive power compensation devices;
- reconfigure the network by opening and closing circuit open points.

1.4.2   Virtual power plant

Distributed energy resources (DER) such as micro-generation, distributed generation, electric vehicles and energy storage devices are becoming more numerous due to the many initiatives to de-carbonize the power sector. DERs are too small and too numerous to be treated in a similar way to central generators and are often connected to the network on a 'connect-and-forget' basis. The concept of a Virtual Power Plant (VPP) is to aggregate many small generators into blocks that can be controlled by the system operator and then their energy output is traded [11]. Through aggregating the DERs into a portfolio, they become visible to the system operator and can be actively controlled. The aggregated output of the VPP is arranged to have similar technical and commercial characteristics as a central generation unit.

The VPP concept allows individual DERs to gain access to and visibility in the energy markets. Furthermore, system operators can benefit from the optimal use of all the available capacity connected to the network.

The size and technological make-up of a VPP portfolio have a significant effect on the benefits of aggregation seen by its participants. For example, fluctuation of wind generation output can lower the value of the energy sold but variation reduces with increasing geographical distance between the wind farms. If a VPP assembles generation across a range of technologies, the variation of the aggregated output of these generators is likely to reduce.

1.4.3   Other initiatives and demonstrations

1.4.3.1   Galvin electricity initiative

The Galvin vision [12, 13] is an initiative that began in 2005 to define and achieve a 'perfect power system'. The perfect power system is defined as:

> *"The perfect power system will ensure absolute and universal availability and energy in the quantity and quality necessary to meet every consumer's needs."*

The philosophy of a perfect power system differs from the way power systems traditionally have been designed and constructed which assumes a given probability of failure to supply customers, measured by a reliability metric, such as Loss of Load Probability (LOLP). Consideration of LOLP shows that a completely reliable power system can only be provided by using an infinite amount of plant at infinite cost.

Some of the attributes of the perfect power system are similar to those of the Smart Grid. For example, in order to achieve a perfect power system, the power system must meet the following goals:

- be smart, self-sensing, secure, self-correcting and self-healing;
- sustain the failure of individual components without interrupting the service;
- be able to focus on regional, specific area needs;
- be able to meet consumer needs at a reasonable cost with minimum resource utilisation and minimal environmental impact;
- enhance quality of life and improve economic productivity.

The development of the perfect power system is based on integrating devices (smart loads, local generation and storage devices), then buildings (building management systems and micro CHP), followed by construction of an integrated distribution system (shared resources and storage) and finally to set up a fully integrated power system (energy optimisation, market systems and integrated operation).

1.4.3.2   IntelliGrid^{SM}

EPRI's IntelliGrid^{SM} initiative [12, 14], which is creating a technical foundation for the Smart Grid, has a vision of a power system that has the following features:

- *is made up of numerous automated transmission and distribution systems, all operating in a coordinated, efficient and reliable manner;*
- *handles emergency conditions with 'self-healing' actions and is responsive to energy-market and utility business enterprise needs;*
- *serves millions of customers and has an intelligent communications infrastructure enabling the timely, secure and adaptable information flow needed to provide reliable and economic power to the evolving digital economy.*

To realise these attributes, an integrated energy and communication systems architecture should first of all be developed. This will be an open standard-based architecture and technologies such as data networking, communication over a wide variety of physical media and embedded computing will be part of it. This architecture will enable the automated monitoring and control of the power delivery system, increase the capacity of the power delivery system, and enhance the performance and connectivity of the end users.

In addition to the proposed communication architecture, the realisation of the IntelliGrid^{SM} will require enabling technologies such as automation, distributed energy resources, storage, power electronic controllers, market tools, and consumer portals. Automation will become widespread in the electrical generation, consumption and delivery systems. Distributed energy resources and storage devices may offer potential solutions to relieve the necessity to strengthen the power delivery system, to facilitate a range of services to consumers and to provide electricity to customers at lower cost, and with higher security, quality, reliability and availability. Power electronic-based controllers can direct power along specific corridors, increase the power transfer capacity of existing assets, help power quality problems and increase the efficient use of power. Market tools will be developed to facilitate the efficient planning for expansion of the power delivery system, effectively allocating risks, and connecting consumers to markets. The consumer portal contains the smart meter that allows price signals, decisions, communication signals and network intelligent requests to flow seamlessly through the two-way portal.

1.4.3.3 Xcel energy's Smart Grid

Xcel Energy's vision [15] of a smart grid includes

> "*a fully network-connected system that identifies all aspects of the power grid and communicates its status and the impact of consumption decisions (including economic, environmental and reliability impacts) to automated decision-making systems on that network.*"

Xcel Energy's Smart Grid implementation involved the development of a number of quick-hit projects. Even though some of these projects were not fully realised, they are listed below as they illustrate different Smart Grid technologies that could be used to build intelligence into the power grid:

1. *Wind Power Storage*: A 1 MW battery energy storage system to demonstrate long-term emission reductions and help to reduce impacts of wind variability.
2. *Neural Networks*: A state-of-the-art system that helps reduce coal slagging and fouling (build-up of hard minerals) of a boiler.
3. *Smart Substation*: Substation automation with new technologies for remote monitoring and then developing an analytics engine that processes data for near real-time decision-making and automated actions.
4. *Smart Distribution Assets*: A system that detects outages and restores them using advanced meter technology.
5. *Smart Outage Management*: Diagnostic software that uses statistics to predict problems in the power distribution system.
6. *Plug-in Hybrid Electric Vehicles*: Investigating vehicle-to-grid technology through field trials.
7. *Consumer Web Portal*: This portal allows customers to program or pre-set their own energy use and automatically control their power consumption based on personal preferences including both energy costs and environmental factors.

1.4.3.4 SCE's Smart Grid

Southern California Edison (SCE)'s Smart Grid strategy encompasses five strategic themes namely, renewable and distributed energy resources integration, grid control and asset optimisation, workforce effectiveness, smart metering, and energy-smart customer solutions [16]. SCE anticipates that these themes will address a broad set of business requirements to better position them to meet current and future power delivery challenges. By 2020, SCE will have 10 million intelligent devices such as smart meters, energy-smart appliances and customer devices, electric vehicles, DERs, inverters and storage technologies that are linked to the grid, providing sensing information and automatically responding to prices/event signals.

SCE has initiated a smart meter connection programme where 5 million meters will be deployed from 2009 to 2012. The main objectives of this programme include adding value through information, and initiating new customer partnerships. The services and information they are going to provide include interval billing, tiered rates and rates based on time of use.

1.5   Overview of the technologies required for the Smart Grid

To fulfil the different requirements of the Smart Grid, the following enabling technologies must be developed and implemented:

1. *Information and communications technologies*: These include:
   (a) two-way communication technologies to provide connectivity between different components in the power system and loads;
   (b) open architectures for plug-and-play of home appliances; electric vehicles and microgeneration;
   (c) communications, and the necessary software and hardware to provide customers with greater information, enable customers to trade in energy markets and enable customers to provide demand-side response;
   (d) software to ensure and maintain the security of information and standards to provide scalability and interoperability of information and communication systems.
   These topics are discussed in Chapters 2–4 of this book.
2. *Sensing, measurement, control and automation technologies*: These include:
   (a) Intelligent Electronic Devices (IED) to provide advanced protective relaying, measurements, fault records and event records for the power system;
   (b) Phasor Measurement Units (PMU) and Wide Area Monitoring, Protection and Control (WAMPAC) to ensure the security of the power system;
   (c) integrated sensors, measurements, control and automation systems and information and communication technologies to provide rapid diagnosis and timely response to any event in different parts of the power system. These will support enhanced asset management and efficient operation of power system components, to help relieve congestion in transmission and distribution circuits and to prevent or minimise potential outages and enable working autonomously when conditions require quick resolution.
   (d) smart appliances, communication, controls and monitors to maximise safety, comfort, convenience, and energy savings of homes;
   (e) smart meters, communication, displays and associated software to allow customers to have greater choice and control over electricity and gas use. They will provide consumers with accurate bills, along with faster and easier supplier switching, to give consumers accurate real-time information on their electricity and gas use and other related information and to enable demand management and demand side participation.
   These topics are discussed in Chapters 5–8.

Table 1.1 Application matrix of different technologies

Application area	Requirement	Information and communications technologies	Sensors, control and automation	Power electronics and energy storage
	Plug-and-play of smart home appliances, electric vehicles, microgeneration	•	•	
	Enabling customers to trade in energy markets	•	•	
	Allowing customers to have greater choice and control over electricity use	•	•	
	Providing consumers with accurate bills, along with faster and easier supplier switching	•	•	
Industries, homes	Giving consumers accurate real-time information on their electricity use and other related information	•	•	
	Enabling integrated management of appliances, electric vehicles (charging and energy storage) and microgeneration	•	•	•
	Enabling demand management and demand side participation	•	•	
Transmission and distribution	Enabling rapid diagnosis and timely response to any event on different part of the power system	•	•	
	Supporting enhanced asset management	•	•	•
	Helping relieve congestion in transmission and distribution circuits and preventing or minimising potential outages	•	•	•
Generation	Supporting system operation by controlling renewable energy sources	•	•	• •
	Enabling long-distance transport and integration of renewable energy sources			• •
	Providing efficient connection of renewable energy sources			• •
	Enabling integration and operation of virtual power plants	•	•	• •
Power system as a whole	Providing greater flexibility, reliability and quality of the power supply system	•	•	•
	Balancing generation and demand in real time	•	•	• •
	Supporting efficient operation of power system components	•	•	• •

3. *Power electronics and energy storage*: These include:
   (a) High Voltage DC (HVDC) transmission and back-to-back schemes and Flexible AC Transmission Systems (FACTS) to enable long distance transport and integration of renewable energy sources;
   (b) different power electronic interfaces and power electronic supporting devices to provide efficient connection of renewable energy sources and energy storage devices;
   (c) series capacitors, Unified Power Flow Controllers (UPFC) and other FACTS devices to provide greater control over power flows in the AC grid;
   (d) HVDC, FACTS and active filters together with integrated communication and control to ensure greater system flexibility, supply reliability and power quality;
   (e) power electronic interfaces and integrated communication and control to support system operations by controlling renewable energy sources, energy storage and consumer loads;
   (f) energy storage to facilitate greater flexibility and reliability of the power system.

These topics are discussed in Chapters 9–12 of this book.
   Table 1.1 shows the application matrix of different technologies.

References

[1] Erinmez, I.A., Bickers, D.O., Wood, G.F. and Hung, W.W. (1999) *NGC Experience with frequency control in England and Wales: provision of frequency response by generator*. IEEE PES Winter Meeting, 31 January–4 February 1999, New York, USA.

[2] Sun, Q., Wu, J., Zhang, Y. *et al.* (2010) *Comparison of the development of Smart Grids in China and the United Kingdom*. IEEE PES Conference on Innovative Smart Grid Technologies Europe, 11–13 October 2010, Gothenburg, Sweden.

[3] European Commission (2006) *European SmartGrids Technology Platform:Vision and Strategy for Europe's Electricity*, http://ec.europa.eu/research/energy/pdf/smartgrids_en.pdf (accessed on 4 August 2011).

[4] Department of Energy and Climate Change, UK, *Smarter Grids: The Opportunity*, December 2009, http://www.decc.gov.uk/assets/decc/what%20we%20do/uk%20energy%20supply/futureelectricitynetworks/1_20091203163757_e_@@_smartergridsopportunity.pdf (accessed on 4 August 2011).

[5] Electricity Networks Strategy Group, *A Smart Grid Routemap*, February 2010, http://www.ensg.gov.uk/assets/ensg_routemap_final.pdf (accessed on 4 August 2011).

[6] Kaplan, S.M., Sissine, F., Abel, A. *et al.* (2009) *Smart Grid: Government Series*, The Capitol Net, Virginia.

[7] U.S. Department of Energy, *Smart Grid System Report*, July 2009, http://www.oe.energy.gov/sites/prod/files/oeprod/DocumentsandMedia/SGSRMain_090707_lowres.pdf (accessed on 4 August 2011).

[8] *A Compendium of Modern Grid Technologies*, July 2009, http://www.netl.doe.gov/smartgrid/referenceshelf/whitepapers/Compendium_of_Technologies_APPROVED_2009_08_18.pdf (accessed on 4 August 2011).

[9] European Commission, *ICT for a Low Carbon Economy: Smart Electricity Distribution Networks*, July 2009, http://ec.europa.eu/information_society/activities/

sustainable_growth/docs/sb_publications/pub_smart_edn_web.pdf (accessed on 4 August 2011)

[10] World Economic Forum (2009) *Accelerating Smart Grid Investments*, http://www.weforum.org/pdf/SlimCity/SmartGrid2009.pdf (accessed on 4 August 2011).

[11] Pudjianto, D., Ramsay, C. and Strbac, G. (2007) Virtual power plant and system integration of distributed energy resources. *Renewable Power Generation, IET*, **1** (1), 10–16.

[12] Gellings, C.W. (2009) *The Smart Grid: Enabling Energy Efficiency and Demand Response*, The Fairmont Press, Lilburn.

[13] Galvin, R., Yeager, K. and Stuller, J. (2009) *Perfect Power: How the Microgrid Revolution Will Unleash Cleaner, Greener, More Abundant Energy*, McGraw-Hill, New York.

[14] *The Integrated Energy and Communication Systems Architecture*, 2004, http://www.epri-intelligrid.com/intelligrid/docs/IECSA_VolumeI.pdf (accessed on 4 August 2011).

[15] *XCel Energy Smart Grid: A White Paper*, March 2007, http://smartgridcity.xcelenergy.com/media/pdf/SmartGridWhitePaper.pdf (accessed on 4 August 2011).

[16] *Southern California Edison Smart Grid Strategy and Roadmap*, 2007, http://asset.sce.com/Documents/Environment%20-%20Smart%20Grid/100712_SCE_SmartGridStrategyandRoadmap.pdf (accessed on 4 August 2011).

Part I

Information and Communication Technologies

2

Data Communication

2.1 Introduction

Data communication systems are essential in any modern power system and their importance will only increase as the Smart Grid develops. As a simple example, a data communication system can be used to send status information from an Intelligent Electronic Device (IED) to a workstation (human–machine interface) for display (see Chapter 6). Any co-ordinated control of the power system relies on effective communications linking a large number of devices.

Figure 2.1 shows a model of a simple point-to-point data communication system in which the communication channel is the path along which data travels as a signal. As can be seen from Figure 2.1, the communication channel could be a dedicated link between the Source and Destination or could be a shared medium.

Using the power system as an example, some possible components of this model and the associated physical devices are listed in Table 2.1.

Communication channels are characterised by their maximum data transfer speed, error rate, delay and communication technology used. Communication requirements for commonly used power systems applications are given in Table 2.2.

2.2 Dedicated and shared communication channels

Certain applications require the transmission of data from one point to another and other uses may require the transmission of data from one point to multiple points. When a secure communication channel is required from one point to another, a dedicated link is used exclusively by the Source and Destination only for their communication. In contrast, when a shared communication channel is used, a message sent by the Source is received by all the devices connected to the shared channel. An address field within the message specifies for whom it is intended. Others simply ignore the message.

Figure 2.2 shows a typical communication network used inside a substation. Each bay has a controller which takes the local measurements (e.g. from current and voltage transformers) and contains the software required for protection and control of the bay primary equipment (e.g. transformers and circuit breakers). These bay controllers are connected to substation control and monitoring equipment (station computer, RTUs) through a star or ring connection

Smart Grid: Technology and Applications, First Edition.
Janaka Ekanayake, Kithsiri Liyanage, Jianzhong Wu, Akihiko Yokoyama and Nick Jenkins.
© 2012 John Wiley & Sons, Ltd. Published 2012 by John Wiley & Sons, Ltd.

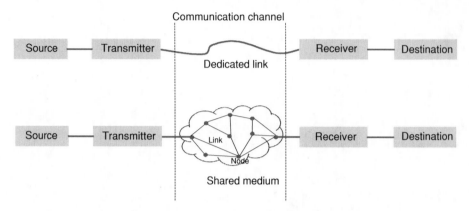

Figure 2.1 Model of simple point-to-point communication system

as shown in Figures 2.2a and 2.2b. In the star connection (Figure 2.2a), each bay controller has a dedicated link to the station computer. In the ring connection (Figure 2.2b), the bay controllers and the station computer are connected through a shared medium to form a Local Area Network (LAN).

Dedicated communication channels are used for differential protection of transmission lines. The communication channel is used to transmit a signal corresponding to the summation of the three line currents (they are added using a summation transformer [2]) at one relaying point to another for comparison with similar signal at that point (see Figure 2.3).

In the differential protection scheme of Figure 2.3, a pilot wire communication or power line carrier may be used. A bit stream from the differential IED is first modulated with a carrier.

Table 2.1 Examples of the physical devices in a power system communication system

Component	Physical device
Source	Voltage transformer
	Current transformer
Transmitter	Remote terminal unit (RTU)
Communication channel	LAN (Ethernet)
Receiver	Network interface card
Destination	Work station with graphic display
	IED for protection and control

Table 2.2 Power system applications and the communication requirements from [1]

Application	Response time required (including latency)	Network topology and communication technology
Protection of transmission circuits	< 20 ms	Dedicated point-to-point links
Protection of distribution circuits	< 100 ms	Circuit switching and packet switching

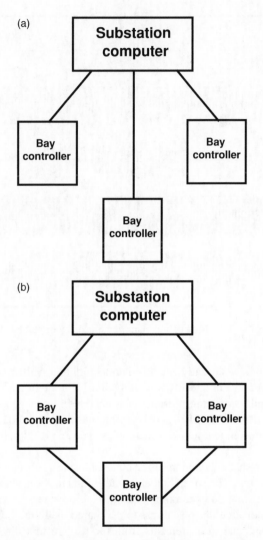

Figure 2.2   Communication within the substation [2]. (a) star connection, (b) ring connection

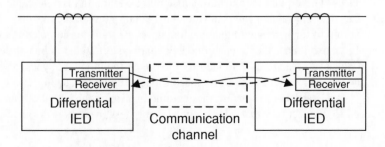

Figure 2.3   Differential relay

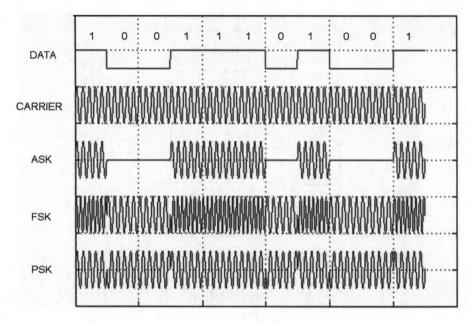

Figure 2.4   Modulation of carrier signals by digital bit streams

The modulation techniques usually employed are: Amplitude Shift Keying (ASK), Frequency Shift Keying (FSK) and Phase Shift Keying (PSK) [1, 3, 4]. In ASK, the carrier frequency remains constant and the digital information is encoded in changes to carrier amplitude, as shown in Figure 2.4. In both FSK and PSK, the carrier amplitude remains constant with changes in the carrier frequency, or the phase of the carrier signal being used to modulate the digital information.

Figure 2.5 shows a shared communication channel inside a substation that uses a multi-drop connection arrangement. The substation computer sends messages addressed to one or more IEDs; each IED takes its turn to communicate with the substation computer to avoid conflicts that could arise due to simultaneous access to the shared medium.

In Figure 2.5, analogue measurements from the CT are first digitised using an encoder (situated inside the IED). The simplest and most widely used method of digitising is Pulse Code Modulation (PCM) where the analogue signal is sampled at regular intervals to produce a series of pulses. These pulses are then quantised by assigning a discrete value as shown in Figure 2.6. This discrete value is then converted to a binary number that results in a bit stream which is subjected to further encoding depending upon the transmission medium used.

Bits may be transmitted either asynchronously, that is, start and stop bits at the beginning and end of a block of data are used to identify the bits in the block, or synchronously, where the sender and receiver maintain the same speed for sending and receiving data using a clock signal [1, 3, 4]. An example of asynchronous communication is the EIA 232 standard[1] which is widely used to connect terminal equipment to computers over short distances. Figure 2.6

---

[1] Formally known as RS 232.

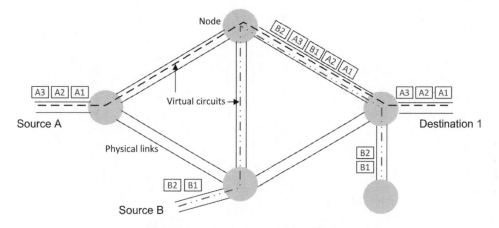

Figure 2.7   Virtual circuit packet switching

important feature of virtual circuit packet switching is that the order of packet delivery is preserved. In other words, packets transmitted from the Source will reach the Destination in the same order as they left the Source. The packets transmitted through several virtual circuits are multiplexed through physical links as shown in Figure 2.7. This allows efficient utilisation of the capacity of a physical link.

Since the path from Source to Destination is set up at the beginning, the need to carry a complete Destination address in each packet is eliminated. This contributes to the reduction of the transmission overhead. Virtual circuit packet switching has the disadvantage that in the case of a node failure, there is complete loss of circuit and the path has to be re-established from the beginning.

2.3.3.2   Datagram packet switching

In datagram packet switching, packets are handled independently (Figure 2.8). Therefore, each packet needs to carry a complete Destination address in it. There is no guarantee that packets belonging to one Source will follow the same path. Consequently, there is no guarantee that packets will arrive at the Destination in the order they were transmitted from the Source. The failure of a node during transmission only affects packets in that node and the communication session will not be disrupted as in the case of virtual circuit packet switching.

2.4   Communication channels

Communication channels run through physical media between a Source and a Destination. In the case of dedicated channels, a single medium, as shown in Figure 2.9, is generally used. Shared communication channels may involve more than one medium, depending on the route the signal travels. A communication channel may be provided through guided media such as a copper cable or optical fibre or through an unguided medium such as a radio link.

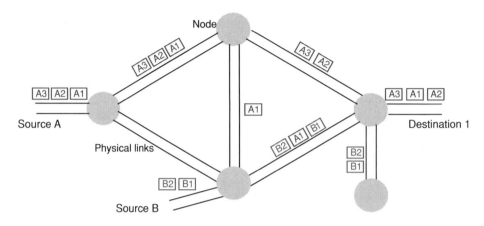

Figure 2.8 Datagram packet switching

The performance of a communication channel is mainly described by the following parameters:

1. *Bandwidth/Bit rate*: Bandwidth is the difference between the upper and lower cut-off frequencies of a communication channel. In an analogue system it is typically measured in Hertz. In digital transmission the term bit rate is often used to express the capacity (some books refer to as bandwidth) of a channel. The bit rate is measured in bits per second (bps).
2. *Attenuation*: As a signal propagates along a communication channel, its amplitude decreases. In long-distance transmission, amplifiers (for analogue signals) and repeaters (for digital signals) are installed at regular intervals to boost attenuated signals. For example, when transmitting digital signals in copper cables, repeaters are required every 10 km, whereas optical fibre can take a signal without significant attenuation over a distance as great as 100 km.
3. *Noise*: In communication, electrical noise is an inherent problem. When digital signals are travelling inside the channel, sometimes noise is sufficient to change the voltage level corresponding to logic '0' to that of logic '1' or vice versa. Noise level is normally described by the Signal to Noise Ratio (SNR) and measured in decibels (dB). SNR is

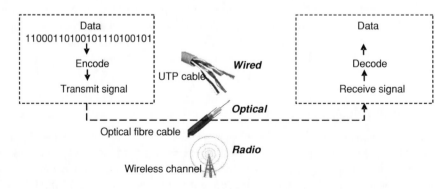

Figure 2.9 Example of data transmission media (wired, optical or radio)

defined by:

$$SNR = 10 \log_{10} \left[\frac{\text{signal power}}{\text{noise power}} \right] \qquad (2.1)$$

For example, if the SNR = 20 dB, from Equation (2.1), it may be seen that the ratio of signal power to noise power is $10^{(20/10)} = 100$.

4. *Signal propagation delay*: The finite time delay that it takes for a signal to propagate from Source to Destination is known as propagation delay. In a communication channel both the media and repeaters (see Chapter 3) that are used to amplify and reconstruct the incoming signals cause delays. As some of the Smart Grid applications require real-time low latency communication capabilities, it is important to consider the propagation delay of a channel.

2.4.1 Wired communication

2.4.1.1 Open wire

Early telephone circuits used two open wire lines and this technology is still used in some countries. With an open wire circuit, care should be taken to avoid cross-coupling of electrical signals with adjacent circuits.

Power Line Carrier (PLC) that uses the power line as a physical communication media could also be considered an open wire communication system. It offers the possibility of sending data simultaneously with electricity over the same medium. PLC uses a Line Matching Unit (LMU) to inject signals into a high voltage transmission or distribution line as shown in Figure 2.10. The injected signal is prevented from spreading to other parts of the power network by line traps.

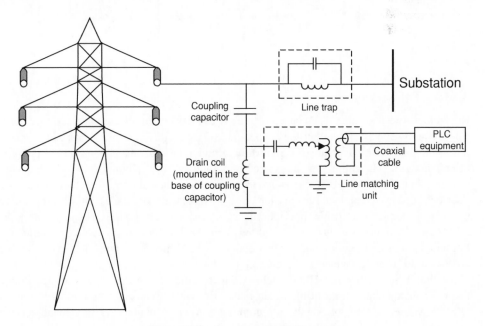

Figure 2.10   Structure of a PLC system

Example 2.1

In Figure 2.10, the PLC equipment uses a carrier frequency of 100 kHz. If the value of the inductance in the line trap is 0.25 mH, what is the value of the capacitance required?

Answer

Figure 2.11   Line trap

Impedance of the line trap,

$$Z = \frac{j\omega L \times 1/j\omega C}{j\omega L + 1/j\omega C} = \frac{L/C}{j[\omega L - 1/\omega C]}$$

At parallel resonance, that is when $\omega L = 1/\omega C$, $Z \to \infty$ and the line trap will block any frequency component at the frequency $f = 1/2\pi\sqrt{LC}$
When $f = 100$ kHz and $L = 0.25$ mH,

$$C = \frac{1}{(2\pi f)^2 L} = \frac{1}{(2\pi \times 100 \times 10^3)^2 \times 0.25 \times 10^{-3}}$$

$$= 10\,\text{nF}$$

2.4.1.2   Twisted pair

Unshielded twisted pair (UTP) cables are used extensively in telecommunication circuits. These usually consist of two twisted copper cables, each with an outer PVC or plastic insulator. Depending on the data rate (which is influenced by the cable material as well as the types of connectors), UTP cables are categorised into a number of categories (or CAT). For voice transmission, Category 1 UTP cables are used. However, they are not suitable for data transmission. For low speed data transmission up to 4 Mbps, Category 2 UTP cables can be used. Category 3, 4 and 5 UTP cables provide data transmission rate of up to 10 Mbps, 16 Mbps and 100 Mbps. Category 5 is the most commonly used UTP cable type in data communication. For applications that require much higher data rates, Category 5e, 6, 6e, 7 and 7a UTP cables are available. They can support data rates up to 1.2 Gbps.

A Digital Subscriber Line (DSL) allows the transmission of data over ordinary copper twisted pair telephone lines at high data rates to provide broadband services. A variant of DSL, Asymmetric DSL or ADSL, is a commonly used technology for broadband services to homes.

2.4.1.3   Coaxial cables

In coaxial cables a shielded copper wire is used as the communication medium. The outer coaxial conductor provides effective shielding from external interference and also reduces losses due to radiation and skin effects. Bit rates up to 10 Mbps are possible over several metres.

Example 2.2

According to Shannon's capacity formula [5], the maximum channel capacity in bps is given by $B \log_2[1 + (\text{signal power})/(\text{noise power})]$, where B is the bandwidth of a channel in Hz. Compare the maximum channel capacity of twisted copper and coaxial cables. For copper cable, the bandwidth is 250 kHz and the SNR is 20 dB. For coaxial cable, the bandwidth is 150 MHz and the SNR is 22 dB.

Answer

For twisted Cu cable:

$$\frac{\text{signal power}}{\text{noise power}} = 10^{(20/10)} = 100$$

Maximum channel capacity $= 250 \times 10^3 \times \log_2 [1 + 100] \approx 1.7$ Mbps.
For coaxial cable:

$$\frac{\text{signal power}}{\text{noise power}} = 10^{(22/10)} = 158$$

Maximum channel capacity $= 150 \times 10^6 \times \log_2 [1 + 158] \approx 1.1$ Gbps.

2.4.2   *Optical fibre*

Optical fibre transmission is used both inside substations and for long-distance transmission of data. Optical fibres are often embedded in the stranded conductors of the shield (ground) wires of overhead lines. These cables are known as OPtical Ground Wires (OPGW). As shown in Figure 2.12a, an OPGW cable contains a tubular structure with one or more optical fibres in it, surrounded by layers of steel and aluminium wire. Optical fibres may be wrapped around the phase conductors or sometimes a standalone cable, an all-dielectric self-supporting (ADSS) cable, is used.

As shown in Figure 2.12b, an optical fibre consists of three components: core, cladding and buffer. The thin glass centre of the fibre where the light travels is called the core. The outer optical material surrounding the core that reflects the light back into the core is called the

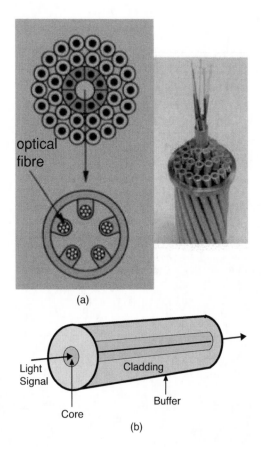

Figure 2.12 Optical fibres. *Source:* OPGW cable picture, courtesy of TEPCO, Japan

cladding. In order to protect the optical surface from moisture and damage, it is coated with a layer of buffer coating.

Compared to other communication media, fibre optic cables have a much greater bandwidth. They are less susceptible to signal degradation than copper wire and their weight is less than a copper cable. Unlike electrical signals in copper wires, light signals from one fibre do not interfere with those of other fibres in the same cable. Further, optical fibre transmission is immune to external electromagnetic interference (EMI). This is important in power system applications since data transmission through the electrically hostile area of a substation is required.

The main disadvantages of optical fibre transmission include the cost, the special termination requirements and its vulnerability (it is more fragile than coaxial cable).

Figure 2.13 shows the path of a light signal travelling inside an optical fibre. A light signal from the optical source is first incident on surface A and then refracted inside the core. The signal is then incident on the surface between the core and cladding. The subsequent path depends on the incident angle, θ_1 [6].

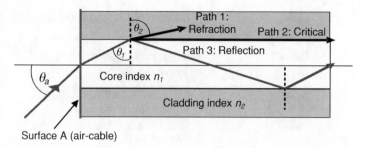

Figure 2.13 Principle of fibre optics

By applying the law of refraction at surface A, the following equation can be obtained:

$$n_0 \sin \theta_a = n_1 \sin \left(\frac{\pi}{2} - \theta_1 \right) = n_1 \cos \theta_1 \tag{2.2}$$

where θ_a is the angle of acceptance n_1, n_2 and n_0 are the refractive indices of core, cladding and air.

Since, for the free space $n_0 = 1$, from Equation (2.2):

$$\sin \theta_a = n_1 \cos \theta_1 \tag{2.3}$$

At the core–cladding surface:

- If the light signal takes path 1 $n_1 \sin \theta_1 = n_2 \sin \theta_2$ and the signal will not be received at the Destination.
- For critical reflection (path 2), $n_1 \sin \theta_1 = n_2 \sin \pi /2 = n_2$.
- Path 3 shows reflection (essential to guide a signal along the core of an optical fibre cable from Source to Destination):

$$n_1 \sin \theta_1 > n_2 \tag{2.4}$$

From Equation (2.4):

$$1 - \cos^2 \theta_1 > \left(\frac{n_2}{n_1} \right)^2 \tag{2.5}$$

By substituting for $\cos \theta_1$ from Equation (2.3), the following equation is obtained from Equation (2.5):

$$\left(\frac{\sin \theta_a}{n_1} \right)^2 + \left(\frac{n_2}{n_1} \right)^2 < 1$$

$$\therefore \theta_a < \sin^{-1} \left(n_1^2 - n_2^2 \right)^{1/2}$$

$\sin \theta_a = \left(n_1^2 - n_2^2 \right)^{1/2}$ is referred to as the numerical aperture of the optical fibre.

Depending on the core diameter, there may be multiple transmission paths or a single transmission path within the core of a fibre. Optical fibre cables having core diameters of

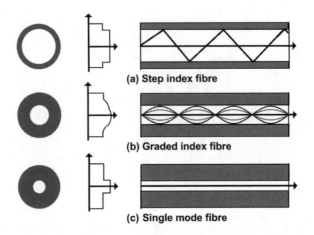

Figure 2.14   Light travel in different fibres

50–400 μm reflect light entering the core from different angles, establish multiple paths (as shown in Figures 2.14a and b) and are called multimode fibres. On the other hand, fibre with a much smaller core diameter, 5–10 μm, supports a single transmission path. This is called single mode fibre (Figure 2.14c). Single mode fibres have advantages such as low dispersion, low noise, and can carry signals at much higher speeds than multimode fibres. Therefore, they are preferred for long-distance applications.

Commonly used multimode fibres are:

1. *Step index fibre*: This cable has a specific index of refraction for the core and the cladding. It is the cheapest type of cable. Its large core diameter allows efficient coupling to incoherent light sources such as Light Emitting Diodes (LED). Different rays emitted by the light source travel along paths of different lengths as shown in Figure 2.14a. As the light travels in different paths, it appears at the output end at different times.
2. *Graded index fibre*: In graded index fibre, rays of light follow sinusoidal paths as shown in Figure 2.14b. Although the paths are different lengths, all the light reaches the end of the fibre at the same time.

Typical parameters of step index, graded index and single mode fibres are given in Table 2.3 [7].

Table 2.3   Typical parameters of step index, graded index and single mode fibres

	Core diameter (μm)	Cladding diameter (μm)	Bandwidth (MHz)	Attenuation (dB km^{-1})	Numerical aperture
Step index	50–400	125–500	6–50	2.6–50	0.16–0.5
Graded index	50–100	100–150	300–3000	2–10	0.2–0.3
Single mode	5–10	125	> 500	2–5	0.08–0.15

Example 2.3

A step-index multimode fibre has a core of refractive index 1.5 and cladding of refractive index 1.485.

1. What is the maximum allowable angle of acceptance for refraction on core–cladding surface?
2. If the length of the fibre is 500 m, what is the difference of distance of travel between the longest and shortest signal path?

Answer

From Equation (2.6)

$$\theta_a < \sin^{-1}\left(n_1^2 - n_2^2\right)^{1/2}$$

$$< \sin^{-1}\left(1.5^2 - 1.485^2\right)^{1/2}$$

$$< 12.2°$$

The maximum allowable angle of acceptance for reflection is 12.2°, shown in Figure 2.15.

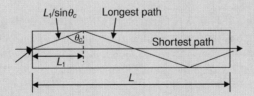

Figure 2.15   Longest and shortest signal paths

Shortest distance $= L = 500$ m
Longest distance $= L/\sin\theta_c$

$$\sin\theta_a = n_1 \cos\theta_c$$
$$\sin(12.1°) = 1.5\cos\theta_c$$
$$\theta_c = 82°$$

Therefore, the longest distance $= 500/\sin 82° = 505$ m

The difference between the shortest and longest distance of travel $= 5$ m.

2.4.3   Radio communication

The substations of power networks are often widely distributed and far from the control centre. For such long distances, the use of copper wire or fibre optics is costly. Radio links provide an alternative for communication between the Control Centre and substations. Even though radio communication cannot provide the bandwidth offered by wired technology, the reliability, performance and running costs of radio networks have improved considerably over the past few years, making it an attractive option.

Radio communication may be multipoint or point-to-point, operating typically either at UHF frequencies (between 300 MHz and 3 GHz) or microwave frequencies (between 3 and 30 GHz).

2.4.3.1   Ultra high frequency

UHF radio represents an attractive choice for applications where the required bandwidth is relatively low and where the communication end-points are widespread over harsh terrain. It uses frequencies between 300 MHz and 3 GHz. Unlike microwave radio, UHF does not require a line of sight between the Source and Destination. The maximum distance between the Source and Destination depends on the size of the antennae and is likely to be about 10–30 km with a bandwidth up to 192 kbps.

2.4.3.2   Microwave radio

Microwave radio operates at frequencies above 3 GHz, offering high channel capacities and transmission data rates. Microwave radio is commonly used in long-distance communication systems. Parabolic antennas are mounted on masts and towers at the Source to send a beam to another antenna situated at the Destination, tens of kilometres away. Microwave radio offers capacity ranging from a few Mbps to hundreds of Mbps. However, the capacity of transmission over a microwave radio is proportional to the frequency used, thus, the higher the frequency, the bigger the transmission capacity but the shorter the transmission distance. Microwave radio requires a line of sight between the Source and Destination, hence, high masts are required. In case of long-distance communications, the installation of tall radio masts will be the major cost of microwave radio.

2.4.4   Cellular mobile communication

Cellular mobile technology offers communication between moving objects. To make this possible, a service area is divided into small regions called cells. Each cell contains an antenna which is controlled by a Mobile Telephone Switching Office (MTSO) as shown in Figure 2.16. In a cellular network, the MTSO ensures the continuation of communication when a mobile device moves from one cell to another. For example, assume that a mobile user in cell A communicates through the antenna located in the base station of cell A. If the user moves out of range of cell A and into the range of cell B, the MTSO automatically assigns a communication channel in cell B to the user without interrupting the communication session.

2.4.5   Satellite communication

Satellites have been used for many years for telecommunication networks and have also been adopted for Supervisory Control And Data Acquisition (SCADA) systems. A satellite

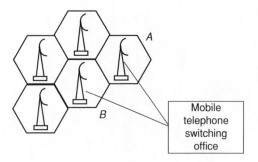

Figure 2.16   A cellular system

communication network can be considered as a microwave network with a satellite acting as a repeater.

2.4.5.1   Geostationary orbit satellite communication

Currently, many satellites that are in operation are placed in Geostationary Orbit (GEO). A GEO satellite or GEOS is typically at 35,786 km above the equator and its revolution around the Earth is synchronised with the Earth's rotation. The high altitude of a GEO satellite allows communications from it to cover approximately one-third of the Earth's surface [8].

Even though GEO satellite-based communication offers technical advantages in long-distance communication, it still presents some drawbacks. They include:

1. The challenge of transmitting and detecting the signal over the long distance between the satellite and the user.
2. The large distance travelled by the signal from the Source to reach the Destination results in an end-to-end delay or propagation delay of about 250 ms.

2.4.5.2   Low earth orbiting (LEO) satellite communication

LEO satellites are positioned 200–3000 km above the Earth, which reduces the propagation delay considerably. In addition to the low delay, the proximity of the satellite to the Earth makes the signal easily detectable even in bad weather.

LEO satellite-based communication technology offers a set of intrinsic advantages such as: rapid connection for packet data, asynchronous dial-up data availability, reliable network services, and relatively reduced overall infrastructure support requirements when compared to GEO. In addition, LEO satellite-based communication channels can support protocols such as TCP/IP since they support packet-oriented communication with relatively low latency.

2.5   Layered architecture and protocols

There is a set of rules, referred to as a protocol that defines how to perform a task associated with a data communication process. A collection of such protocols that work together to

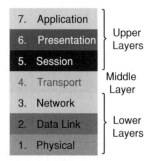

Figure 2.17   ISO/OSI reference model

support data communication between Source and Destination is called a protocol stack. In a stack, protocols are arranged in layers. Arranging protocols in layers can also be viewed as decomposing the complex task of information exchange into smaller sub-tasks that are mostly independent from each other. Therefore, a protocol can be changed or modified without affecting the other protocols of the overall communication task or other protocols in the stack.

2.5.1   The ISO/OSI model

The 'Open System Interconnection model' developed by the International Standard Organisation (ISO/OSI) [2, 4] is a protocol architecture which consists of seven layers that describes the tasks associated with moving information from Source to Destination through a communication network. The seven-layer stack of the ISO/OSI reference model is shown in Figure 2.17.

In this model, apart from the Physical layer, all the other layers use the services of the layer immediately below them. For instance, the Application layer uses the services of the Presentation layer to ensure that information sent from the Application layer of one system would be readable by the Application layer of another system using different data representation formats.

As shown in Figure 2.17, the seven layers in this model can be viewed as consisting of three groups of protocol sets. The Upper Layers present an interface to the user. They are Application[3] orientated and are not concerned with how data is delivered to Applications. The Lower Layers are more concerned with data transmission and are Application independent. Usually service providers offer services defined in the Lower Layers. The Middle Layer separates Application orientated Upper Layers from transmission-orientated Lower Layers.

---

[3] In Chapters 2, 3, 7 and 8, the word 'Application' is used to refer to computer software that helps the user to perform specific tasks.

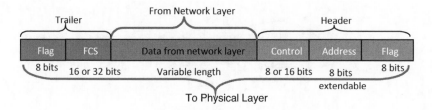

Figure 2.18 Frame structure of the HDLC protocol

2.5.1.1 The physical layer

The Physical layer is responsible for transmitting data (in the form of bits) from one node to the next as a signal over a transmission channel. It deals with the signalling methods, data encoding, bandwidth and multiplexing required for data transmission depending on the communication channel used.

2.5.1.2 The data link layer

The Data Link layer prepares data blocks usually referred to as frames for transmission, takes care of the synchronisation of data transmission at both Source and Destination, and resolves issues related to accessing the medium shared by multiple users. Figure 2.18 shows the frame structure of High-Level Data Link Control (HDLC)[4] protocol. In it the bits are organized in a sequence starting from the header on the right.

The main functions of the Data Link layer include:

1. *Framing*: As shown in Figure 2.17, data received from the Network layer is encapsulated in a frame by adding header and trailer information. The header usually contains information such as flag, a unique bit sequence to indicate the beginning of a frame, Source and Destination addresses and various control information required to ensure the correct delivery of the frame. The trailer usually contains information that is required to detect errors, and a unique bit sequence to indicate the end of a frame.
2. *Error control*: For the purpose of error control, each frame is numbered using a finite number of bits referred to as sequence number (Frame Check Sequence – FCS). The sequence number allows the data link layer to decide which frames are to be transmitted, which frames have reached their Destination with errors and need retransmission and which frames are to be acknowledged as having been received without errors.
3. *Medium Access Control (MAC)*: In networks where a medium is shared by multiple users, it is required to determine data to decide which user should transmit when. This is referred to as access control of a shared medium and is one function of the Data Link layer.
4. *Physical addressing*: A unique address assigned to a physical device is used to identify the correct Destination or nodes to which to deliver a frame. Usually this address is called the MAC address assigned by the network adaptor manufacturer.

---

[4] A widely used synchronous data link protocol which forms the basis for many other data link control protocols.

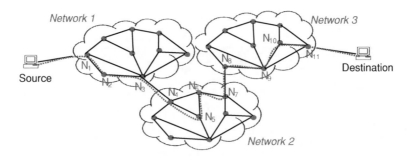

Figure 2.19   Route from Source to Destination through multiple networks

2.5.1.3   The network layer

This layer is responsible for the delivery of data packets from the Source to Destination across the communication network. For example, as shown in Figure 2.19, the path between Source and Destination Hosts traverses three networks which could be of three different types. One of the main functions of the Network layer is to use packet switching techniques (as described in Section 2.3) to switch data at each node.

Other functions of the Network layer include:

1. *Traffic control*: Traffic control attempts to control the number of packets that are in a network at a time to prevent the network from becoming a bottleneck. Traffic control can be achieved through flow control, congestion control and routing. Flow control regulates the rate of the data flow between two nodes. Congestion, which occurs if the number of queued packets at a node grows large, is controlled by regulating the number of packets in the entire network. Thus, congestion control occurs in multiple nodes. Routing which facilitates the updating route calculation considering loads at nodes is also used to relieve congestion.
2. *Addressing*: The Network layer is the lowest layer that involves moving data from Source to Destination as shown in Figure 2.20.

The Network layer uses an addressing scheme for the delivery of packets from Source to Destination across multiple networks. These addresses of Source and Destination enable the delivery of packets over networks. The IP address is a good example of an address used in Network layer and will be described in detailed in Section 2.5.2.

2.5.1.4   The transport layer

The Transport layer accepts data from the Session layer and segments the data for transport across the network. Generally, the Transport layer is responsible for making sure that the data is delivered error-free and in the proper sequence. Flow control, which is the process of managing the rate of data transmission between two nodes to prevent a fast sender from outrunning a slow receiver [3], generally occurs at the Transport layer. The Transport layer also provides the acknowledgement of the successful data transmission.

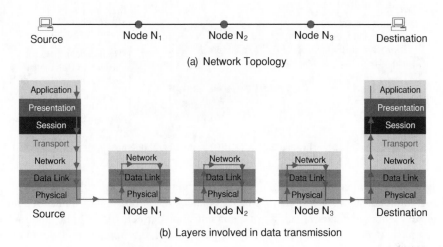

(a) Network Topology

(b) Layers involved in data transmission

Figure 2.20   End-to-end transmission of data

2.5.1.5   The session layer

The Session layer establishes, manages, and terminates communication sessions. Communication sessions consist of service requests and service responses that occur between Applications located in different network devices. These requests and responses are coordinated by protocols such as the Session Announcement Protocol (SAP) [10] and L2TP [11] implemented at the Session layer.

2.5.1.6   The presentation layer

The Presentation layer provides a variety of coding and conversion functions that are applied to Application layer data. These functions ensure that information sent from the Application layer of one system would be readable by the Application layer of another system. Some examples of Presentation layer coding and conversion schemes include common data representation formats, conversion of character representation formats, common data compression schemes and common data encryption schemes.

Common data representation formats, or the use of standard image, sound and video formats, enable the interchange of Application data between different types of computer systems. Conversion schemes are used to exchange information with systems by using different text and data representations, such as EBCDIC and ASCII. Standard data compression schemes enable data that is compressed at the Source to be properly decompressed at the Destination.

2.5.1.7   The application layer

The Application layer is the OSI layer closest to the end user, which means that both the OSI Application layer and the user interact directly with the Application.

This layer interacts with Applications that implement a communication component. Such Applications fall outside the scope of the OSI model. Application layer functions typically

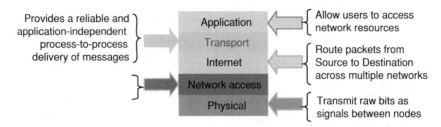

Figure 2.21   TCP/IP protocol architecture

include identifying communication partners, determining resource availability, and synchronising communication. When identifying communication partners, the Application layer determines the identity and availability of communication partners for an Application with data to be transmitted. When determining resource availability, the Application layer must decide whether sufficient network resources for the requested communication are available. In synchronised communication, all communication between Applications requires cooperation, that is also managed by the Application layer.

2.5.2   TCP/IP

The Transmission Control Protocol (TCP)/Internet Protocol (IP) or TCP/IP is the most widely used protocol architecture today. It is a result of a project called Advanced Research Projects Agency Network (ARPANET) funded by the Defense Advanced Research Project Agency (DARPA) in the early 1970s. The TCP/IP protocol architecture used in the Internet evolved out of ARPANET.

Five layers, as shown in Figure 2.21, are defined in the TCP/IP architecture. They are: Physical layer, Network access layer, Internet layer, Transport layer and Application layer [9].

The role of the Physical layer of the TCP/IP is identical to that of the Physical layer in the ISO/OSI reference model. It deals with the specifications of electrical and mechanical aspects of interfaces and transmission media. It is also responsible for encoding data into signals, defining data rate and the synchronisation of bits.

The Network layer is responsible for providing an error-free channel for the Internet layer. Its functions include [3]: encapsulation of IP packets coming from the Internet layer into frames, frame synchronisation, error detection and correction, logical link control, providing flow and error control, media access control, physical addressing (MAC addressing), LAN switching, data packet queuing or scheduling algorithms, and IP address to/from physical address resolution.

The main responsibility of the Internet layer is routing packets from Source to Destination. Identifying Hosts uniquely and universally is essential for routing packets as datagrams[5] across networks. The Internet layer uses an identifier called the IP address to identify devices connected to a network. There are two versions, IPv4 and IPv6, of IP addressing currently in use. IP version 4 (IPv4) is still the most commonly used. Since its introduction in 1998 with

---

[5] A datagram is a variable-length packet consisting of header and data (see Section 2.5.1.2). The header contains the information required for routing and delivery [3].

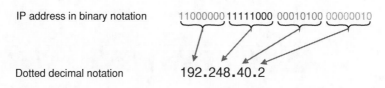

IP address in binary notation   11000000 11111000 00010100 00000010

Dotted decimal notation   192.248.40.2

Figure 2.22   IPv4 address notations

the publication of IETF RFC2373 [12] and RFC2460 [13], IP version 6 is becoming more widely used in the internet.

The Transport layer is represented by the Transmission Control Protocol (TCP). TCP creates a virtual circuit between the Source to Destination and is responsible for sending all datagrams generated by the Source. TCP establishes a transmission by informing the Destination that there are more data to be transmitted and terminates the connection only when all the datagrams have been transmitted.

The Application layer in TCP/IP is a combination of session, presentation and Application layers of the OSI model. Some of the protocols associated with the Application layer of TCP/IP are: Domain Name Server (DNS), File Transfer Protocol (FTP), electronic mail protocols such as Simple Mail Transfer Protocol (SMTP), Hypertext Transfer Protocol (HTTP) and Uniform Resource Locator (URL). More information about these is found in [3, 4].

2.5.2.1   IP version 4

IPv4 addresses are 32 bits long. Usually these are displayed as a sequence of 4 octets[6] with space between octets to make the addresses more readable. In order to make the address compact, a notation called dotted decimal notation is commonly used to represent an IPv4 address as shown in Figure 2.22.

IPv4 addressing has two architectures called classful and classless addressing. Classful addressing is the concept used initially and is still in use widely. Classless addressing was introduced in the mid-1990s and is expected to supersede classful addressing:

1. *Classful addressing*: In classful addressing, the 32-bit address space is divided into five classes called A, B, C, D and E. They are defined according to the bit pattern of the most significant octet as shown in Figure 2.23. The IPv4 classful addressing scheme recognises the class of a network by looking at the first byte (octet) of the address as shown in Figure 2.23. The 32-bit address of a Host has two parts, namely Network ID and Host ID. Network ID is common for all Hosts in a given network and Host ID is used to identify Hosts uniquely. Network ID is used to transport data across networks and Host ID is used to deliver data to a particular Host within a network. As shown in Figure 2.23, the Class A network can theoretically have 2^{24} (16,777,216) Hosts since it uses 24 bits to represent the Host ID. Likewise, the Class B and C networks can theoretically have 2^{16} (65,536) and 2^{8} (256) Hosts. In Class D a datagram is directed to multiple uses. Classful addressing could

---

[6] An octet is a sequence of 8 bits. Even though 8 bits are generally called a byte, there is no standard definition for a byte.

First byte
Second byte
Third byte
Fourth byte
Network ID
Host ID
Class A
0 - 127
0 - 255
0 - 255
0 - 255
Network ID
Host ID
Class B
128 - 191
0 - 255
0 - 255
0 - 255
Network ID
Host ID
Class C
192 - 223
0 - 255
0 - 255
0 - 255
Multicast address
Class D
224 - 239
0 - 255
0 - 255
0 - 255
Reserved for future use
Class E
240 - 247
0 - 255
0 - 255
0 - 255

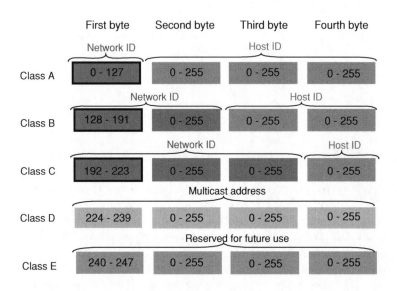

Figure 2.23 IPv4 classful addressing

lead to a waste of IP addresses. For instance, not many organisations will be able to use Class A address space completely.

2. *Classless addressing*: Classless addressing allows variable length blocks to assign only the required number of IP addressed to an organisation. Slash notation, which indicates the number of bits used for Network ID of the address, is used to identify network address as shown below:
   (a) 192.248.40.0/24 – Only the first three bytes as shown under Class C in Figure 2.23 are used for Network ID. Here slash 24 stands for three bytes ($3 \times 8 = 24$ bits)
   (b) 192.248.40.252/30 – The first three bytes and six leading bits from the fourth byte are used for Network ID. Here slash 30 stands for the first three bytes plus six bits from the last byte ($3 \times 8 + 6 = 30$ bits).

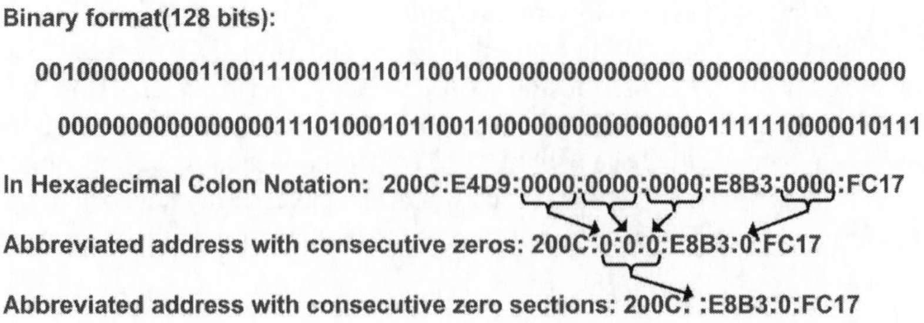

Binary format(128 bits):
001000000000110011100100110110010000000000000000 0000000000000000
0000000000000000111010001011001100000000000000001111110000010111
In Hexadecimal Colon Notation: 200C:E4D9:0000:0000:0000:E8B3:0000:FC17
Abbreviated address with consecutive zeros: 200C:0:0:0:E8B3:0:FC17
Abbreviated address with consecutive zero sections: 200C: :E8B3:0:FC17

Figure 2.24 IPv6 address

2.5.2.2 IP version 6

IP version 6 also known as IP Next Generation (IPng) is a 128-bit addressing scheme. Therefore, it provides a much bigger address space compared to that of IPv4. The main advantages provided by IPv6 include:

1. Internet Protocol Security (IPsec) is mandatory for IPv6. It is a protocol suite for securing Internet Protocol (IP) communications by authenticating and encrypting each IP packet of a communication session.
2. Support for jumbograms which can be as large as 4, 294, 967, 295 ($2^{32} - 1$) octets. In contrast, IPv4 supports datagrams up to 65, 535 ($2^{16} - 1$) octets.

An IPv6 address is denoted using a format called hexadecimal colon notation as shown in Figure 2.24.

References

[1] Aggarwal, R. and Moore, P. (1993) Digital communications for protection. *Power Engineering Journal*, **7**(6), 281–287.
[2] *Network Protection and Automation Guide: Protective Relays, Measurements and Control*, May 2011, Alstom Grid, Available from http://www.alstom.com/grid/NPAG/ on request.
[3] Forouzan, B.A. (2007) *Data Communication and Networking*, McGraw-Hill, New York.
[4] Halsall, F. (1995) *Data Communications, Computer Networks and Open Systems*, Addison-Wesley, Wokingham.
[5] Stallings, W. (2005) *Wireless Communications and Networks*, Pearson Educational International, New Jersey.
[6] Lacy, E. (1982) *Fiber Optics*, Prentice-Hall International, London.
[7] Senior, J. (1992) *Optical Fiber Communications: Principles and Practice*, Prentice-Hall International, London.
[8] Hu, Y. and Li, V. (2001) Satellite-based internet: A tutorial. *IEEE Communications Magazine*, **39**(3), 154–162.
[9] Stallings, W. (2007) *Data and Computer Communications*, Pearson Prentice-Hall, New Jersey.
[10] Handley, M., Perkins, C. and Whelan, E. *Session Announcement Protocol*, IETF RFC 2974, http://www.rfc-editor.org/rfc/rfc2974.txt (accessed on 4 August 2011).
[11] Townsley, W., Valencia, A., Rubens, A. *et al. Layer Two Tunneling Protocol L2TP*, IETF RFC 2661, http://www.rfc-editor.org/rfc/rfc2661.txt (accessed on 4 August 2011).
[12] Hinden, R. and Deering, S. *IP Version 6 Addressing Architecture*, http://www.rfc-editor.org/rfc/rfc2373.txt (accessed on 4 August 2011).
[13] Deering, S. and Hinden, R. *Internet Protocol, Version 6, (IPv6) Specification*, http://www.rfc-editor.org/rfc/rfc2460.txt (accessed on 4 August 2011).

3

Communication Technologies for the Smart Grid

3.1 Introduction

The communication infrastructure of a power system typically consists of SCADA systems with dedicated communication channels to and from the System Control Centre and a Wide Area Network (WAN). Some long-established power utilities may have private telephone networks and other legacy communication systems. The SCADA systems connect all the major power system operational facilities, that is, the central generating stations, the transmission grid substations and the primary distribution substations to the System Control Centre. The WAN is used for corporate business and market operations. These form the core communication networks of the traditional power system. However, in the Smart Grid, it is expected that these two elements of communication infrastructure will merge into a Utility WAN.

An essential development of the Smart Grid (Figure 3.1) is to extend communication throughout the distribution system and to establish two-way communications with customers through Neighbourhood Area Networks (NANs) covering the areas served by distribution substations. Customers' premises will have Home Area Networks (HANs). The interface of the Home and Neighbourhood Area Networks will be through a smart meter or smart interfacing device.

The various communication sub-networks that will make up the Smart Grid employ different technologies (Table 3.1) and a key challenge is how they can be integrated effectively.

In the ISO/OSI reference model, discussed in Chapter 2, the upper layers deal with Applications of the data irrespective of its actual delivery mechanism while the lower layers look after delivery of information irrespective of its Application. In this chapter, communication technologies that are associated with the lower three layers of the ISO/OSI reference model are discussed.

Smart Grid: Technology and Applications, First Edition.
Janaka Ekanayake, Kithsiri Liyanage, Jianzhong Wu, Akihiko Yokoyama and Nick Jenkins.
© 2012 John Wiley & Sons, Ltd. Published 2012 by John Wiley & Sons, Ltd.

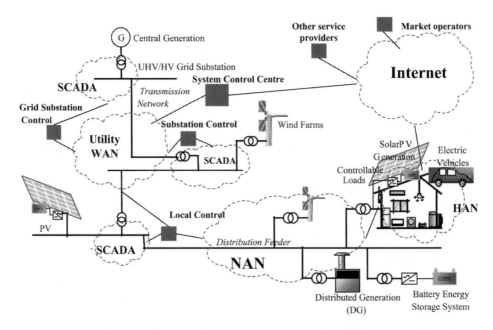

Figure 3.1   Possible communication infrastructure for the Smart Grid

3.2   Communication technologies

3.2.1   IEEE 802 series

IEEE 802 is a family of standards that were developed to support Local Area Networks (LANs). For the Smart Grid illustrated in Figure 3.1, IEEE 802 standards are applicable to LANs in SCADA systems, NANs around the distribution networks and HANs in consumers' premises. Table 3.2 shows the different IEEE 802 standards applicable to commonly used communication technologies with their frequency band, bandwidth, bit rate and range.

Figure 3.2 shows how the IEEE 802 architecture relates to the lowest two layers of the ISO/OSI reference model [3, 4]. It shows how two LANs may be connected through a Bridge (see Table 3.4). Such a connection is common in many organisations which have multiple LANs. A packet from the Source enters the Logical Link Control (LLC) sublayer which acts as

Table 3.1   Technologies used in different sub-networks

Sub-network	Communication technologies
HAN	Ethernet, Wireless Ethernet, Power Line Carrier (PLC), Broadband over Power Line (BPL), ZigBee
NAN	PLC, BPL, Metro Ethernet, Digital Subscriber Line (DSL), EDGE, High Speed Packet Access (HSPA), Universal Mobile Telecommunications System (UMTS), Long Term Evolution (LTE), WiMax, Frame Relay
WAN	Multi Protocol Label Switching (MPLS), WiMax, LTE, Frame Relay

Table 3.2 Different technologies specified under IEEE 802 [1, 2]

Protocol	Description	Frequency band	Bandwidth	Bit rates	Range
IEEE 802.3	Ethernet – Section 3.2.1.1			See Table 3.2	
IEEE 802.4	Token bus – This is a LAN with each device in the network logically connected as a ring [4].			1, 5 and 10 Mbps	
IEEE 802.11a	Wireless LAN (WiFi) – Section 3.2.1.2	5 GHz	20 MHz	6, 9, 12, 18, 24, 36, 48, 54 Mbps	Indoor: 35 m Outdoor: 120 m
IEEE 802.11b		2.4 GHz	20 MHz	1, 2, 5.5, 11 Mbps	Indoor: 38 m Outdoor: 140 m
IEEE 802.11g		2.4 GHz	20 and 40 MHz	1, 2, 6, 9, 12, 18, 24, 36, 48, 54 Mbps	Indoor: 38 m Outdoor: 140 m
IEEE 802.11n		2.4 and 5 GHz	20 and 40 MHz	varies between 6.5 to 300 Mbps	Indoor: 70 m Outdoor: 250 m
IEEE 802.15.1	Bluetooth – Section 3.2.1.3	2.4 GHz		1–3 Mbps	Class 1 – 1 m Class 2 – 10 m Class 3 – 100 m
IEEE 802.15.4	This standard applies to low duty-cycle communication. It specifies and controls physical and MAC layers – Section 3.2.1.4	868.3 MHz 902–928 MHz 2400–2483.5 MHz	600 kHz 2000 kHz 5000 kHz	20 kbps 40 kbps 250 kbps	
IEEE 802.16	WiMAX (Worldwide Inter-operability for Microwave Access) – Section 3.2.1.5.	2–66 GHz	1.25, 5, 10 and 20 MHz	75 Mbps (for fixed and 15 Mbps (for mobile)	50 km

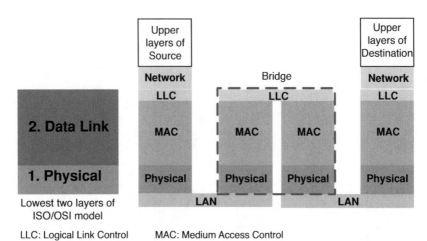

LLC: Logical Link Control   MAC: Medium Access Control

Figure 3.2   IEEE 802 architecture [4]

Table 3.3   Ethernet Standards and their coverage [4]

Standard	Medium	Data rate	Maximum segment length (m)
10BASE5	Thick coaxial cable (50 Ω)		500
10BASE2	Thin coaxial cable (50 Ω)		185
10BASE-T	Unshielded twisted pair (UTP)	10 Mbps	100
10BASE-FP	62.5/125 μm multimode optical fibre pair		500
100BASE-TX	2 pairs Category 5 UTP		100
100BASE-FX	2 optical fibres	100 Mbps	100
100BASE-T4	4 pair Category 3,4, or 5 UTP		100
	10 μm single mode fibre		5000
1000BASE-LX	50 μm multimode fibre		550
	62.5 μm multimode fibre		550
1000BASE-SX	50 μm multimode fibre		550
	62.5 μm multimode fibre	1000 Mbps	220
1000BASE-T	Category 5, 5e, 6 and 7 UTP		100
1000BASE-CX	Shielded cable		25
1000BASE ZX	Single mode fibre		~70000
10GBASE-S	50 μm multimode fibre	1 Gbps	300
	62.5 μm multimode fibre		26–82
10GBASE-L	Single mode fibre		10000
10GBASE-E	Single mode fibre		40000
	Single mode fibre		10000
10GBASSE-LX4	50 μm multimode fibre		300
	62.5 μm multimode fibre		300

an interface between the network layer and the MAC sublayer. The LLC sublayer is defined by IEEE 802.2 and provides multiplexing mechanisms, flow control and error control. The packet then passes into the MAC sublayer. At the MAC sublayer, a header and a trailer (depending on the LAN which the packet is entering) are added to the packet. Then it goes through the physical layer and the communication channel and reaches the Bridge. At the MAC layer of the Bridge, the header and trailer are removed and the original packet is recovered and passes to the LLC sublayer of the Bridge. Then the packet is processed (by adding an appropriate header and trailer) for the LAN to which it is forwarded (to the Destination) by the MAC layer. This use of a Bridge is essential as different LANs use different frame lengths and speeds. For example, IEEE 802.3 uses a frame of 1500 bytes whereas IEEE 802.4 uses a frame of 8191 bytes [4].

3.2.1.1   Ethernet

Ethernet has become the most widely used network technology for wired LANs due to its simplicity, ease of maintenance, ability to incorporate new technologies and reliability. It has a low cost of installation and is easy to upgrade. It is a frame-based communication technology that is based on IEEE 802.3. Its baseband is defined in a number of standards such as 10BASE5, 10BASE2, 10BASE-T, 1BASE5, 100BASE-T, and so on. The first number, that is, 1, 10 and 100, indicates the data rate in Mbps. The last number or letter indicates the maximum length of the cable and type of the cable as defined in Table 3.3. An Ethernet LAN consists of all or some of the devices shown in Table 3.4.

Table 3.4   Data transmission devices for Ethernets

Device	Description
Repeater	A Repeater has two ports. Once it receives a signal, this is amplified to eliminate any distortion (which has been introduced when it was travelling through the communication channel) and forwarded to the output port. A Repeater works at the physical layer of the ISO/OSI reference model. A Repeater is shown in Figure 3.3. Note that even though the input is digital, it can be weak and distorted as shown in Figure 3.3.
Hub	A Hub is a multiport repeater which links multiple Ethernet devices. A Hub passes the incoming signal to all the devices connected to it (Figure 3.4).
Bridge	A Bridge has two ports and operates in the data link layer. It transmits the incoming frame only if the channel to its destination is free or the frame is a broadcast frame. Each port on a bridge supports a full duplex operation.
Switch	A Switch is a multiport bridge. Unlike a hub, a Switch will not broadcast frames across the entire network (unless it is a broadcast frame), it only sends the frame to the intended ports (Figure 3.5).
Router	A Router is used as a gateway between a LAN and a WAN. A Router makes intelligent decisions on how to route traffic. Routing protocols are composed of different algorithms that direct the way routers move traffic. A Router operates in the network layer.

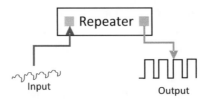

Figure 3.3 Transmission path of a repeater

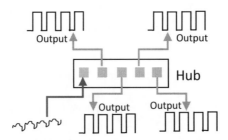

Figure 3.4 Transmission path of a hub

Ethernet uses a shared medium where more than one device tries to use the common medium. This causes collisions of frames transmitted by multiple hosts. The issue of collision is handled by a protocol called Carrier Sense Multiple Access/Collision Detect (CSMA/CD). A set of hosts connected to a network in such a way that simultaneous transmission by two hosts in the set leads to collision, creates a collision domain.

Ethernet LANs also carry broadcast frames defined by the addressing of layer 3 of the ISO/OSI model. The domain to which these broadcast frames reach is called the broadcast domain.

Network performance under heavy traffic conditions is affected by the way collision domains and broadcast domains are located within the network. Therefore isolating them properly is vital to maintain peak performance of LANs. A typical Ethernet LAN-based network, showing the collision and broadcast domains, is shown in Figure 3.6. It is important to note that Switches,

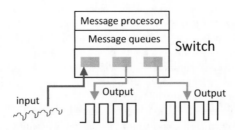

Figure 3.5 Transmission path of a switch

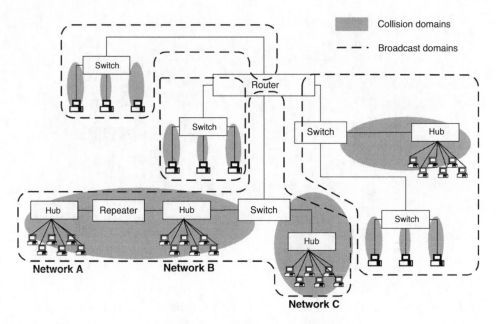

Figure 3.6   A typical Ethernet LAN

Bridges and Routers limit collision domains and Routers also limit broadcast domains. In the network shown in Figure 3.6, a packet leaving from Network A may collide with a packet leaving from Network B but not one leaving from Network C. This is because Networks A and B and Network C are connected through a Switch which limits the collision domains as shown. However, broadcast messages to Networks A, B and C may collide.

Example 3.1

It is necessary to connect two sets of IEDs to the station bus (see Chapter 6). Propose a suitable design.

Answer:

Two sets of IEDs could be connected as in Figure 3.7. Devices A and B could be Hubs or Switches. If Hubs are used, then they will create a collision domain across the station bus as shown by the light shaded area round Devices A and B in Figure 3.7. Therefore, they are not suitable for critical applications that use status information and measurements from IEDs. If this is the case, Devices A and B should be Switches which limits the collision domains within the dark shaded areas round Devices A and B.

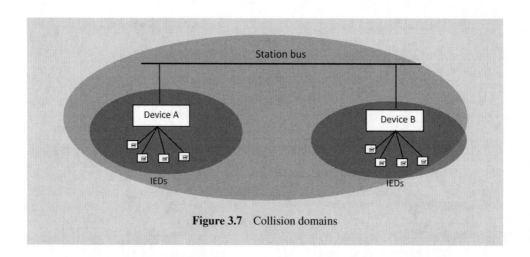

Figure 3.7 Collision domains

3.2.1.2 Wireless LANs

IEEE 802.11 describes the Wireless LAN (WLAN) standard. The interoperability of IEEE 802.11 devices is certified by the Wi-Fi Alliance.[1] Wireless LANs consist of the following components:

1. *Station*: This describes any device that communicates over a WLAN, for example, a notebook computer, or mobile phones that support WiFi. In ad-hoc networks these devices can communicate between themselves by creating a mesh network as shown in Figure 3.8a. Such a collection of stations forming an ad-hoc network is called an Independent Basic Service Set (Independent BSS or IBSS).
2. *Access points (AP)*: When an AP is present in a network, it allows one station to communicate with another through it. It needs twice the bandwidth that is required if the same communication is done directly between communicating stations. However, there are benefits of having an AP in a network. APs make the system scalable and allow wired connection to other networks. Also APs buffer the traffic when that station is operating in a very low power state. When an AP is present in the network, as shown in Figure 3.8b, the collection of stations is called an Infrastructure BSS.
3. *Distribution system (DS)*: A Distribution System interconnects multiple Infrastructure BSSs through their APs as shown in Figure 3.9. It facilitates communication between APs, forwarding traffic from one BSS to another and the movement of mobile stations among BSSs. Such a set of Infrastructure BSSs is called an Extended Service Set (ESS).

The 802.11 family of Wireless LANs use the CSMA/CA protocol to access the transmission medium. They can have multiple physical layer options identified by 802.11a/b/g/n/ as described in Table 3.2. A typical 802.11 application in the Smart Grid is shown in Figure 3.10.

---

[1] http://www.wi-fi.org/

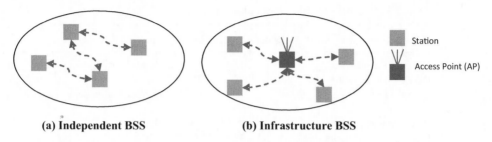

(a) Independent BSS   **(b) Infrastructure BSS**

Figure 3.8   BSS architectures of Wireless LANs

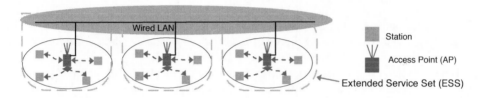

Figure 3.9   Distribution system

3.2.1.3 Bluetooth

Bluetooth, defined by IEEE standard 802.15.1, is a wireless LAN technology designed to connect mobile or fixed devices using low-power, short-distance radio transmission. It was originally conceived as a wireless alternative to EIA 232 data cables. Bluetooth has its classic version currently defined by Bluetooth 3.0+HS and the recently introduced low-energy

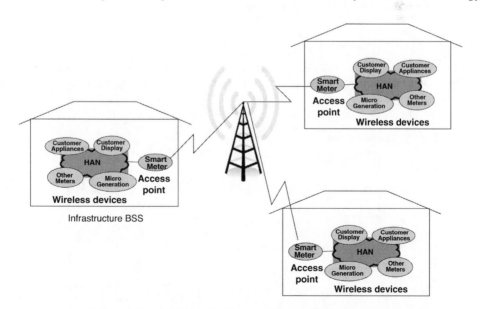

Figure 3.10   IEEE 802.11 WLAN application in the Smart Grid

Table 3.5 Technical specifications of Bluetooth

Specification	Bluetooth 3.0+HS	Bluetooth 4.0
Distance/Range	Up to 100 m	150 m
Over the air data rate	1–3 Mbps	1 Mbps
Active slaves	7	Implementation dependent
Voice capable	Yes	No
Power consumption	100 mW	10 mW
Peak current consumption	< 30 mA	< 20 mA
Topology	Piconet, Scatternet	star, point-to-point

version, Bluetooth 4.0. This new version is designed for applications which require low power consumption and transfer small pieces of data with low latency. Table 3.5 gives a limited technical description of these two Bluetooth standards.

Bluetooth defines two network architectures called Piconet and Scatternet. Piconet is a Bluetooth network consisting of a master device and up to seven slave devices as shown in Figure 3.11. More devices can exist in synchronisation with the master but are not be able to participate in communication concurrently. A slave in such state is said to be in a parked state. A device in a parked state can move to an active state if the number of slaves in the Piconet falls below seven. Piconets can be interconnected through a Bridge which could be a slave for one Piconet and master for another Piconet or slave for two Piconets that are interconnected as shown in Figures 3.11a and 3.11b. Such an interconnected set of Piconets is called a Scatternet.

Two types of Bluetooth links can be created for data transfer. They are Synchronous Connection Orientated (SCO) link and Asynchronous Connectionless Link (ACL). SCO is used when timely delivery is more important than error-free delivery. On the other hand, ACL is used when error-free delivery is more important than timely delivery.

3.2.1.4   ZigBee and 6LoWPAN

ZigBee and 6LoWPAN are two communication technologies built on IEEE 802.15.4. This is a low data rate wireless networking standard. Currently this standard is the most popular protocol for a Wireless Public Area Networks (WPAN) due to its low power consumption,

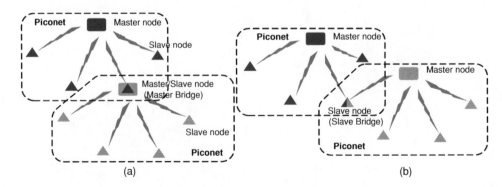

(a)   (b)

Figure 3.11   Piconet and Scatternet

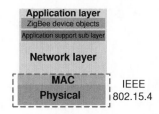

Figure 3.12   Protocol architecture of Zigbee [2]

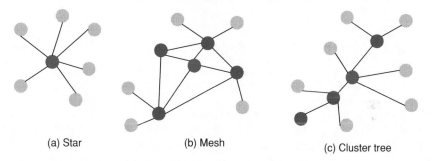

(a) Star   (b) Mesh   (c) Cluster tree

Figure 3.13   ZigBee network architectures

high flexibility in networking and low cost. It creates an ad-hoc self-organising network by interconnecting fixed, portable and moving devices.

The protocol architecture of a ZigBee device is shown in Figure 3.12. As shown in Figure 3.12, the lower two layers are defined by IEEE 802.15.4 standards. Application support and Network layer protocols for a ZigBee network are defined by the ZigBee Alliance.

A ZigBee device can be a Full Function Device (FFD) or a Reduced Function Device (RFD). A network will have at least one FFD, operating as the WPAN coordinator. The FFD can operate in three modes: a coordinator, a router or an end device. An RFD can operate only as an end device. An FFD can talk to other FFDs and RFDs, whereas an RFD can only talk to an FFD. An RFD could be a light switch or a sensor which communicates with a controller. ZigBee networks can have star, mesh or cluster tree architecture, as shown in Figure 3.13.

6LoWPAN is a protocol which enables IPv6 packets to be carried over low power WPAN. The minimum transmission unit for an IPv6 packet is 1280 octets. However, the maximum MAC frame size defined by IEEE 802.15 is 127 bytes. Therefore, in order to implement the connection between the MAC layer and IPv6 network layer in 6LoWPAN, an adaptation layer between the MAC layer and the network layer is placed, as shown in Figure 3.14.

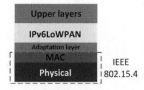

Figure 3.14   6LoWPAN network architecture [2]

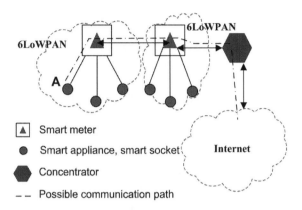

☐ Smart meter
● Smart appliance, smart socket
⬡ Concentrator
– – Possible communication path

Figure 3.15   Communication over a 6LoWPAN network

When an RFD in a 6LoWPAN wants to send a data packet to an IP-enabled device outside the 6LoWPAN domain, it first sends the packet to an FFD in the same WPAN. An FFD which act as a Router in 6LoWPAN will forward the data packet hop by hop to the 6LoWPAN gateway. The 6LoWPAN gateway that connects to the 6LoWPAN with the IPv6 domain will then forward the packet to the destination IP-enabled device by using the IP address. An example is shown in Figure 3.15 where the signal path from Device A to a device in the Internet is shown.

Example 3.2

Figure 3.16 shows a part of a smart metering network. It is proposed to use ZigBee for the communication network. Considering the distances given, propose a suitable ZigBee communication network for this scheme. In the proposed configuration, do the smart meters act as an FFD or as an RFD?

Answer

The typical maximum range of ZigBee is 30 m. So smart meters at houses A and F could connect directly to the concentrator. The smart meter in house A could act as a repeater to the smart meter in house B. Similarly the smart meter in house F could act as a Repeater to the smart meter in house E. The smart meter in house E could act as a Repeater to the smart meter in house D. Since house C is far away from both houses B and D, a Repeater is required to connect that house to the smart meter in house B or D. However, in a communication network, redundancy is added in case the communication system in one smart meter fails. So the final network with redundancy would look like Figure 3.17. In this configuration, as each smart meter communicates with another smart meter as well as the appliances and other devices in its HAN, it acts as an FFD.

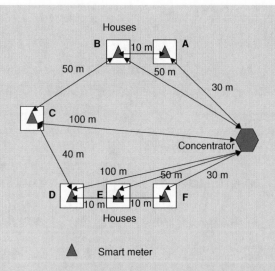

Figure 3.16   Figure for Example 3.2

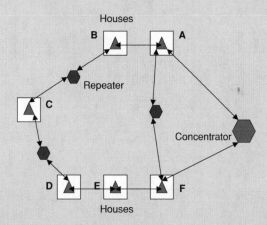

Figure 3.17   ZigBee communication network for smart metering

Example 3.3

Figure 3.18 shows a part of a smart metering network. Each smart meter sends 3 bytes of data every second. The network is designed such that if one link from a concentrator fails, another link will transmit the data. There are 250 000 smart meters in the network and there are 1000 concentrators. What is the average data rate through a data link between the concentrator and the data company?

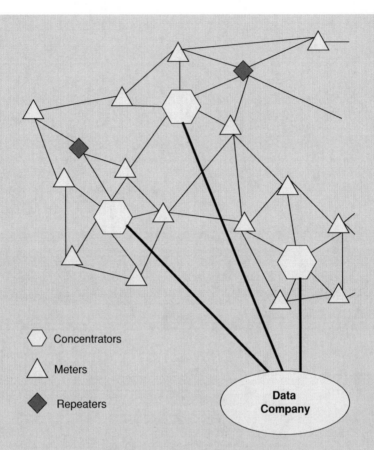

Figure 3.18   Figure for Example 3.3

Answer

On average, the number of smart meters connected to each concentrator $= 250\,000/1000 = 250$.

For the worst case analysis it could be assumed that all 250 smart meters send data at the same time. So the data transfer rate through the link between the concentrator and data company is:

$$= 250 \times 3 \text{ bytes/s}$$

If one link fails, another link should be able to transfer the data on that link. So the data rate per link should be 1.5 kbps.

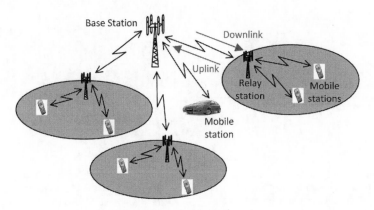

Figure 3.19 WiMax network

3.2.1.5 WiMax

Worldwide Interoperability for Microwave Access (WiMax) is a wireless technology which conforms with the IEEE 802.16 standard. It provides both fixed and mobile connectivity using a technique called Orthogonal Frequency Division Multiple Access (OFDMA). A typical WiMax installation is shown in Figure 3.19.

The coverage of WiMax extends up to 50 km with peak data rates of 75 Mbps for fixed connections and up to 15 Mbps for mobile connections. These data rates are expected to increase beyond 200 Mbps in Uplink and 300 Mbps in Downlink with the release of IEEE 802.16m-2011. It is optimised to support mobile devices moving up to a speed of 10 km/h. Even though it supports vehicles moving up to 120 km/h, its performance degrades with the vehicle speed. It has the capability to maintain connection with stations moving at up to 350 km/h [5].

3.2.2 Mobile communications

Mobile communication systems were designed initially to carry voice only. The standard that has enabled this technology is GSM (Global System for Mobile Communications). As an add-on data service to GSM technology, the General Packet Radio Service (GPRS) was developed. GPRS uses the existing GSM network and adds two new packet-switching network elements: the GGSN (Gateway GPRS Support Node) and the SGSN (Serving GPRS Support Node).

In December 1998, the European Telecommunications Standards Institute, the Association of Radio Industries and Businesses/Telecommunication Technology Committee of Japan, the China Communications Standards Association, the Alliance for Telecommunications Industry Solutions (North America) and the Telecommunications Technology Association (South Korea) launched a project called the 3rd Generation Partnership Project (3GPP). The aim of the 3GPP project was to develop a 3rd Generation mobile systems (3G) based on GSM, GPRS and EDGE (Enhanced Data Rates for GSM Evolution). The project was built on data communication rather than voice. This project rapidly evolved to provide many different technologies

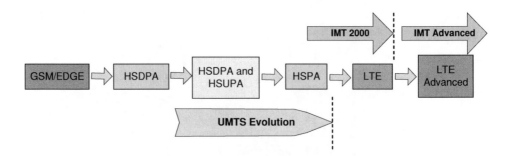

EDGE: Enhanced Data Rates for GSM Evolution   HSPA: High Speed Packet Access
GSM: Global System for Mobile Communications   IMT: International Mobile Telecommunications
HSDPA: High Speed Downlink Packet Access   UMTS: Universal Mobile Telecommunication System
HSUPA: High Speed Uplink Packet Access

Figure 3.20   Evolution of the 3GPP family

as shown in Figure 3.20. The data rates of the different technologies that evolved under 3GPP are shown in Table 3.6.

LTE is a competing technology to WiMax and supports user mobility up to 350 km/h, coverage up to 100 km, channel bandwidth up to 100 MHz with spectral efficiency of the Downlink 30 bps/Hz and the Uplink 15 bps/Hz. LTE has the advantage that it can support seamless connection to existing networks, such as GSM and UMTS as shown in Figure 3.21.

3.2.3   Multi protocol label switching

Multi Protocol Label Switching (MPLS) is a packet forwarding technique capable of providing a Virtual Private Network (VPN) service to users over public networks or the internet. VPN provides the high quality of service and security required by Applications such as that associated with critical assets. Some anticipated Applications of point-to-point VPNs based on MPLS include Remote Terminal Unit (RTU) networks and backbone network to the System Control Centre. MPLS-based VPN is an attractive solution for wide area connectivity due to its relatively low cost and ability to be implemented rapidly using the existing network resources.

Table 3.6   Peak data rates of the 3GPP family

Technology	Peak data rates (Mbps)	
	Uplink	Downlink
GSM/EDGE	0.5	1.6
HSDPA and HSUPA	5.76	14.4
HSPA	22	56
LTE	75	300
LTE-Advanced	500	1000

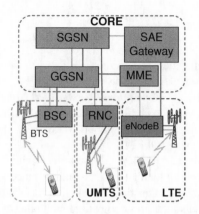

SSGN: Serving GPRS Support Node
GGSN: Gateway GPRS Support
SAE: System Architecture Evolution
MME: Mobility Management Entity
BSC: Base Station Controller
BTS: Base Transceiver Station
RNC: Radio Network Controller
eNodeB: Equivalent to a base station

Figure 3.21 LTE connecting to legacy networks

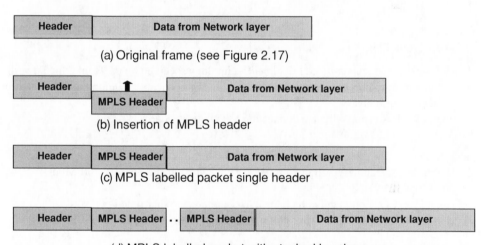

Figure 3.22 MPLS packet

MPLS works by attaching labels to data packets received from the Network layer as shown in Figure 3.22. A MPLS header consists of four fields namely: the 20-bit label field, the 3-bit experimental or class of service field, the stack bit and the 8-bit time to live field as shown in Figure 3.23. In MPLS, when a packet is forwarded, the label is sent with it to the next node. At that node, the label is used to determine the next hop. The old label is replaced with a new label and the packet is forwarded to its next hop.

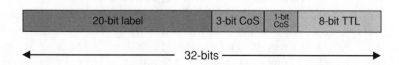

Figure 3.23 MPLS header

3.2.4   Power line communication

3.2.4.1   IEEE P1901

Under the sponsorship of the IEEE Communication Society, the IEEE P1901 working group was formed in 2005 with the remit to develop a standard for high speed (> 100 Mbps at the physical layer) communication devices via electric power lines, the so-called Broadband over Power Line (BPL) devices. This project is devoted to producing a standard for BPL networks. The in-home and access protocol under IEEE P1901 will support MAC layer and Physical layers that use orthogonal frequency multiplexing (OFDM).

The standard which was in draft form at the end of 2010 will use transmission frequencies below 100 MHz and support BPL devices used for the first-mile/last-mile connections as well as BPL devices used in buildings for LANs and other data distribution. It ensures that the EMC limits set by national regulators are met so that it is compatible with other wireless and telecommunications systems.

3.2.4.2   HomePlug

HomePlug is a non-standardized broadband technology specified by the HomePlug Powerline Alliance, whose members are major companies in communication equipment manufacturing and in the power industry.

HomePlug Powerline Alliance defines the following standards:

1. HomePlug 1.0: connects devices in homes (1–10 Mbps).
2. HomePlug AV and AV2: transmits HDTV and VoIP in the home – 200 Mbps (AV) and 600 Mbps (AV2).
3. HomePlug CC: Command and Control to complement other functions.
4. HomePlug BPL: still a working group addressing last-mile broadband (IEEE P1901).

The transmission technology, OFDM used by HomePlug, is specially tailored for use in the power line environments. It uses 84 equally spaced subcarriers in the frequency band between 4.5 and 21 MHz. Impulsive noise events common in the power line environment are overcome by means of forward error correction and data interleaving.

3.3   Standards for information exchange

3.3.1   Standards for smart metering

Smart meters (see Chapter 5) may be used in various ways and these lead to different require-ments for the metering communication system. Automated meter reading (AMR) requires only occasional transmission of recorded energy data (perhaps once a month) while advanced metering infrastructure (AMI) requires frequent bi-directional communication (perhaps every 30 minutes). The use of smart meters to support Smart Grid operation of the distribution network has not yet been implemented widely but is likely to place severe demands on the communication system.

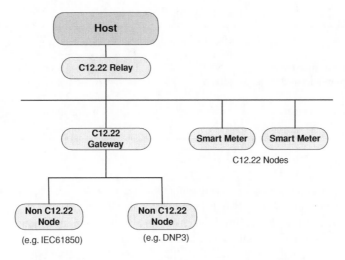

Figure 3.24 Basic ANSI C12.22 smart metering architecture

IEC 62056 and ANSI C12.22 are two sets of standards that describe open communication systems for smart meters. IEC 62056 defines the Transport and Application layers for smart metering under a set of specifications called COSEM (Companion Specification for Energy Metering). ANSI C12.22 specifies the sending and receiving of meter data to and from external systems. It can be used over any communication network. The architecture defined in ANSI C12.22 to transport messages is shown in Figure 3.24.

3.3.2 Modbus

Modbus is a messaging protocol in the Application layer and provides communication between devices connected over several buses and networks. It can be implemented through Ethernet or using asynchronous serial transmission over EIA 232, EIA 422, EIA 485 and optical fibres. Of these, the most common implementation is Modbus over EIA485. Figure 3.25 shows how the

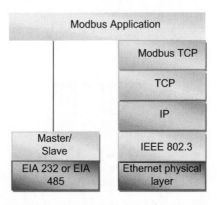

Figure 3.25 Modbus protocol stack [6]

Table 3.7   Comparison of two Modbus implementations [7]

	Modbus over EIA 485	Modbus over Ethernet
Maximum speed	115 kbps	10/100 Mbps/1Gbp
Maximum distance without repeaters	1300 m	UTP 100 m Fibre 2000 m
Maximum number of devices	1 master and 31 slaves	64 input/output scanning, no limit for others

Modbus Application layer connects with other OSI layers. Table 3.7 compares the performance of Modbus implementations in two different physical layers.

Modbus over EIA 485 is used extensively in substation automation. As shown in Figure 2.5, in a substation, multiple devices share a common link.

Communication on a Modbus over EIA 485 is started by a Master with a query to a Slave. The Slave which is constantly monitoring the network for queries will recognise only the queries addressed to it and will respond either by performing an action (setting a value) or by returning a response. Only the Master can initiate a query. The Master can address individual Slaves, or, using a broadcast address, can initiate a broadcast message to all Slaves.

3.3.3   DNP3

DNP3 (Distributed Network Protocol) is a set of communication protocols developed for communications between various types of data acquisition and control equipment. It plays a crucial role in SCADA systems, where it is used by Control Centres, RTUs and IEDs. DNP3 has recently been adopted as an IEEE standard 1815–2010 [8].

DNP3 has five layers as shown in Figure 3.26. The DNP User layer can take analogue and binary inputs and output analogue and binary signals. A Master DNP3 station sends requests and typically the Slave DNP3 stations respond to these requests. However, a Slave DNP3 station may also transmit a message without a request. The DNP3 Physical layer most commonly uses serial communication protocols such as EIA 232 or EIA 485. Recently applications of DNP3 over an Ethernet connection can be found.

Figure 3.26   DNP protocol stack [8]

Table 3.8   Structure of the standard IEC 61850 [9]

Part	Description
1	Introduction and overview
2	Glossary
3	General requirements such as quality requirements, environmental conditions including EMI (electromagnetic interferences) immunity and auxiliary services.
4	Defines engineering requirements such as parameter types (system, process and functional), engineering tools (system specification, system configuration, IED configuration and documentation) and quality assurance.
5	Communication requirements for functions and device models including logical node approach.
6	Substation Configuration Language (SCL) – each device in the substation must provide its configuration according to SCL
7	Communication structure for substation and feeder equipment. This has four parts and defines the information model for substation automation, Application modelled by logical nodes, model of the device database structure and logical node classes and data classes
8 and 9	These provide: • Definitions for mapping objects and services to Manufacturing Mapping Specifications (MMS) and Ethernet • Definitions of mapping GOOSE (Generic Object Oriented Substation Event)[1] messages and GSSE (Generic Substation Status Event) to Ethernet • Mapping(s) of services used for the transmission of sampled analogue values.
10	Conformance testing

[1] Used for fast transmission of substation events, such as commands and alarms as messages.

3.3.4   IEC 61850

IEC 61850 is an open standard for Ethernet communication within substations. It is a function-based standard which ensures interoperability of substation equipment. The functions are divided into:

1. *system support functions*: network management, time synchronisation and physical device self-checking;
2. *system configuration or maintenance functions*: software management, configuration management, settings and test modes;
3. *operational or control functions*: parameter set switching, alarm management and fault record retrievals;
4. *process automation functions*: protection, interlocking and load shedding.

IEC 61850 uses an object model to describe the information available from the different pieces of substation equipment and from the substation controller. The standard contains 10 parts as described in Table 3.8.

   In addition to defining communication protocols, IEC61850 specifies a data structure. As shown in Figure 3.27a, a device model starts by considering a physical device. Then logical

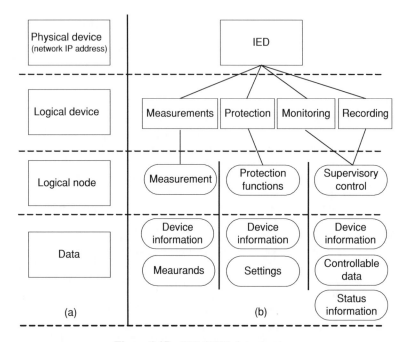

Figure 3.27   IEC 61850 data structure

devices within that device are specified. Each logical device is then mapped into 86 different classes of logical nodes defined in IEC 61850. The names of these nodes are specified. Finally data related to each logical node is specified. An example of an IED is shown in Figure 3.27b. In an IED, logical devices such as measurements, protection, monitoring and recording may be found. Each logical device has multiple logical nodes reflecting their functions. Even though the logical nodes associated with protection logical devices are specified as a single node, they are divided into about 40 logical nodes that include distance, differential, over-current, and so on.

References

[1] IEEE (2011) *IEEE 802 LAN/MAN Standards Committee*, IEEE, March 2011, http://www.ieee802.org/minutes/2011-March/802%20workshop/index.shtml (accessed on 4 August 2011).

[2] The Open Meter Consortium (2009) *Description of Current State-of-the-Art Technologies and Protocols: General Overview of State-of-the-Art Technological Alternatives*, http://www.openmeter.com/files/deliverables/OPEN-Meter%20WP2%20D2.1% 20part1%20v3.0.pdf (accessed on 4 August 2011).

[3] Jeffree, T. (2011) *The IEEE 802.1 Standards*, IEEE, March 2011, http://www.ieee802.org/minutes/2011-March/802%20workshop/index.shtml (accessed on 4 August 2011).

[4] Low, D. (2011) *IEEE 802.3 Ethernet*, IEEE, March 2011, http://www.ieee802. org/minutes/2011-March/802%20workshop/index.shtml (accessed on 4 August 2011).

[5] Cudak, M. (ed.) (2010) *IEEE 802.16m System Requirements*, IEEE 802.16 Task Group M, January 2010, http://ieee802.org/16/tgm/docs/80216m-07_002r10.pdf (accessed on 4 August 2011).

[6] *Modbus Application Protocol Specification V1.1b*, December 2006, http://www.modbus.org/docs/Modbus_Application_Protocol_V1_1b.pdf (accessed on 4 August 2011).

[7] Electrical Installation Organization, *Communication Protocols and Architectures in iPMCC*, http://www.electrical-installation.org/wiki/Communication_protocols_and_architectures_in_iPMCC (accessed on 4 August 2011).

[8] IEEE Power & Energy Society (2010) *IEEE Std 1815TM-2010: IEEE Standard for Electric Power Systems Communications-Distributed Network Protocol (DNP3)*, June.

[9] British Standard Institute (2002–2010) *BS EN/IEC 61850: Communication Networks and Systems for Power Utility Automation*.

4

Information Security for the Smart Grid

4.1 Introduction

The operation of a Smart Grid relies heavily on two-way communication for the exchange of information. Real-time information must flow all the way to and from the large central generators, substations, customer loads and the distributed generators. At present, power system communication systems are usually restricted to central generation and transmission systems with some coverage of high voltage distribution networks. The generation and transmission operators use private communication networks, and the SCADA and ICT systems for the control of the power network are kept separate even from business and commercial applications operated by the same company. Such segregation of the power system communication and control system (using private networks and proprietary control systems) limits access to this critical ICT infrastructure and naturally provides some built-in security against external threats.

With millions of customers becoming part of the Smart Grid, the information and communication infrastructure will use different communication technologies and network architectures that may become vulnerable to theft of data or malicious cyber attacks. Ensuring information security in the Smart Grid is a much more complex task than in conventional power systems because the systems are so extensive and integrated with other networks. Potentially sensitive personal data is transmitted and, in order to control costs, public ICT infrastructure such as the Internet will be used.

Obtaining information about customers' loads could be of interest to unauthorised persons and could infringe the privacy of customers. The ability to gain access to electricity use data and account numbers of customers opens up numerous avenues for fraud. Breaching the security of power system operating information by an unauthorised party has obvious dangers for system operation.

The Smart Grid requires reliable and secure delivery of information in real time. It not only needs *throughput*, the main criterion adopted to describe performance required for common internet traffic [1]. Delays in the delivery of information accurately and safely are less tolerable

Smart Grid: Technology and Applications, First Edition.
Janaka Ekanayake, Kithsiri Liyanage, Jianzhong Wu, Akihiko Yokoyama and Nick Jenkins.
© 2012 John Wiley & Sons, Ltd. Published 2012 by John Wiley & Sons, Ltd.

in the Smart Grid than for much commercial data transmission as the information is required for real-time or near real-time monitoring and control. A lot of monitoring and control information is periodic and contributes to a regular traffic pattern in the communication networks of the Smart Grid, though during power system faults and contingencies there will be a very large number of messages. The length of each message will be short. The pattern of messages passed in the ICT system of the Smart Grid is very different to that of a traditional voice telephone system or the Internet. Any form of interruption resulting from security issues is likely to have serious effects on the reliable and safe operation of the Smart Grid.

Security measures should ensure the following:

1. *Privacy* that only the sender and intended receiver(s) can understand the content of a message.
2. *Integrity* that the message arrives in time at the receiver in exactly the same way it was sent.
3. *Message authentication* that the receiver can be sure of the sender's identity and that the message does not come from an imposter.
4. *Non-repudiation* that a receiver is able to prove that a message came from a specific sender and the sender is unable to deny sending the message.

Providing information security has been a common need of ICT systems since the Internet became the main mode of communication. Thus there are well-established mechanisms to provide information security against possible threats.

4.2   Encryption and decryption

Cryptography has been the most widely used technique to protect information from adversaries. As shown in Figure 4.1, a message to be protected is transformed using a Key that is only known to the Sender and Receiver. The process of transformation is called encryption and the message to be encrypted is called Plain text. The transformed or encrypted message is called Cipher text. At the Receiver, the encrypted message is decrypted.

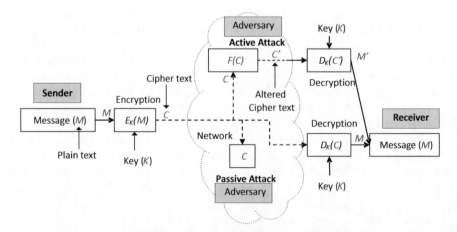

Figure 4.1   Components involved in cryptography and possible threats

Figure 4.1 also shows possible threats to a message. As indicated in Figure 4.1, an intruder can launch a passive or an active attack. A passive attacker may use captured information for malicious purposes. In an active attack, data may be modified on its path or completely new data, may be sent to the Receiver. Passive attacks, though they do not pose an immediate threat, are hard to detect. Active attacks are more destructive but can be detected quickly in most situations.

4.2.1 Symmetric key encryption

In classical encryption both sender and receiver share the same Key. This is called symmetric key encryption.

4.2.1.1 Substitution cipher

Substitution cipher was an early approach based on symmetric Key encryption. In this process, each character is replaced by another character. An example of a mapping in a substitution cipher system is shown below:

Plain text	A	B	C	D	E	F	G	H	I	J	K	L	M	N	O	P	Q	R	S	T	U	V	W	X	Y	Z
Cipher text	W	Y	A	C	Q	G	I	K	M	O	E	S	U	X	Z	B	D	F	H	J	L	N	P	R	T	V

The encryption of message or plain text HELLO THERE will produce KQSSZ JKQFQ as Cipher text. Since a given character is replaced by another fixed character, this system is called a mono-alphabetic substitution. The Key here is the string of 26-characters corresponding to the full alphabet. Substitution cipher systems disguise the characters in the Plain text but preserve the order of characters in the Plain text.

Even though the possible Key combinations, $26! \approx 4 \times 10^{26}$, appear to be large enough that this system feels safe, by using the statistical properties of natural languages, the Cipher text can easily be broken.

The possibility of exploiting the statistical properties of natural languages, due to mono-alphabetic substitution, can be reduced by the use of polyalphabetic substitution, in which each occurrence of a character can have a different substitute. In other words, the relationship between a character in Plain text and a character in Cipher text is one-to-many. For instance, the substituting character corresponding to a character can be made to depend upon the relative position of the character within the Plain text.

4.2.1.2 Transposition cipher

In a transposition cipher the characters in the Plain text are transposed to create the Cipher text. Transposition can be achieved by organising the Plain text into a two-dimensional array and interchanging columns according to a rule defined by a Key. An example of transposition cipher is shown in Figure 4.2. As can be seen, Plain text is first assigned to an array having the same number of columns as the Key. Any unused columns are filled with the letter 'a'. Then each row in the array is rearranged in the alphabetical order of the Key.

Original message: contract to buy three hundred GWhrs from abc generating company

Plain text: contracttobuythree**hundred**dgwhrsfr**omabcgeneratingcompanyaa**

Key: PURCHASE (58624173)

P	U	R	C	H	A	S	E	← Key word
5	8	6	2	4	1	7	3	← Alphabetical order of characters in the Key word
c	o	n	t	r	a	c	t	
t	o	b	u	y	t	h	r	columns rearranged
e	e	h	u	n	d	r	e	according to alphabetical
d	g	w	h	r	s	f	r	order
o	m	a	b	c	g	e	n	
e	r	a	t	i	n	g	c	
o	m	p	a	n	y	a	a	

A	C	E	H	P	R	S	U
1	2	3	4	5	6	7	8
a	t	t	r	c	n	c	o
t	u	r	y	t	b	h	o
d	u	e	n	e	h	r	e
s	h	r	r	d	w	f	g
g	b	n	c	o	a	e	m
n	t	c	i	e	a	g	r
y	a	a	n	o	p	a	m

Cipher text: attrcncoturytbho**duenehre**shrrdwf**ggbncoaem**ntcieagry**aanopam**

Figure 4.2   Transposition cipher example

4.2.1.3   One-time pad

In the one-time pad method a random bit string of the same length as the message is combined using the bit-wise exclusive OR (XOR) with the Plain text. The adversary has no information at all for breaking the Cipher text produced by a one-time pad since every possible Plain text is equally probable.

In this method, both Sender and Receiver have to carry the same random bit string Key of encryption. This is not an easy task as a Key can only be used once and the amount of data that can be transferred with a given Key is limited by the length of the Key.

If a one-time pad Key is used to encrypt more than one message, this can cause vulnerabilities in security. To understand this better, consider a situation where Plain text messages M_1 and M_2 are encrypted with same Key K. Let C_1 and C_2 be the corresponding Cipher texts:

$$C_1 = K \oplus M_1$$

$$C_2 = K \oplus M_2$$

$$Hence\, C_1 \oplus C_2 = M_1 \oplus M_2$$

Therefore, if an adversary gets to know either M_1 or M_2, the other message can be computed easily.

Example 4.1

The word 'HALLOW' is encrypted by one-time pad '00110010 00010000 01111010 00001101 00101001 00010010'. Assume each letter in the plain text is encoded using 8-bit ACSII code. What is the Cipher text that will be generated?

Answer

Table 4.1 Cipher text for Example 4.1

Plain text	H	A	L	L	O	W
Printable characters[1]	0x48	0x41	0x4C	0x4C	0x4F	0x57
ASCII code in binary	01001000	01000001	01001100	01001100	01001111	01010111
One time pad	00110010	00010000	01111010	00001101	00101001	00010010
Cipher text (XOR*)	01111010	01010001	00110110	01000001	01100110	01000101
Corresponding number	0x7A	0x51	0x36	0x41	0x66	0x45
Cipher Text	z	Q	6	A	f	E

Notes: *Bit wise XOR.
[1]This represents letters, digits, punctuation marks, and some symbols in ASCII codes by hexadecimals (http://en.wikipedia.org/wiki/ASCII).

4.2.1.4 Data encryption standard

The Data Encryption Standard (DES) is a typical example of modern cryptography. As a well-engineered encryption standard, it or its variants have been in widespread use since the early 1970s. In DES, blocks of data having 64 bits are treated as units for encryption. Figure 4.3 shows the DES algorithm. As shown in Figure 4.3a, the Plain text is encrypted

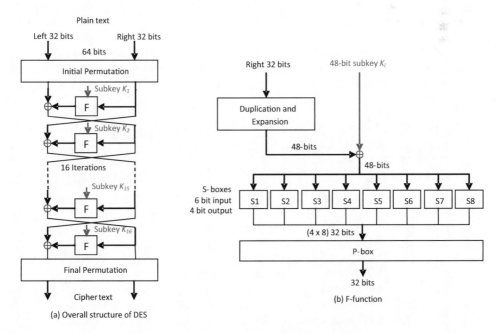

Figure 4.3   The Data Encryption Standard [2]

Table 4.2 Mapping matrix used for initial permutation [2]

58	50	42	34	26	18	10	2
60	52	44	36	28	20	12	4
62	54	46	38	30	22	14	6
64	56	48	40	32	24	16	8
57	49	41	33	25	17	9	1
59	51	43	35	27	19	11	3
61	53	45	37	29	21	13	5
63	55	47	39	31	23	15	7

16 times using 48 bit subkeys ($K_1, \ldots, K_{16}$) generated by a 56-bit Key. The F-function indicated in Figure 4.3a is the Feistel function and shown in Figure 4.3b.

Initial permutation and final permutation are used to rearrange the input and output bits using a mapping method. For example, the matrix shown in Table 4.2 is used to permutate the input. After initial permutation, the 58th bit of the plain text becomes the 1st bit, the 50th bit becomes the 2nd bit, and so on.

The duplication and expansion box is used to expand the right 32 bits of the plain text to 48 bits after initial permutation. The expanded bits are then combined with the 48-bit subkey, K_i, using the XOR operation. The output is then grouped into 6-bit groupings. These groupings are sent through S-boxes which use an address to locate an output using another mapping matrix (see Example 4.2). Each S-box uses different mapping matrices and provides 4-bit output. Each output is then combined to form a 32-bit stream and that goes through a P-box (use a mapping matrix as the initial permutation to rearrange the bits) to form the output of each F-function.

Example 4.2

An S-box receives a 6-bit input 100 111. Its output is determined by the following mapping matrix. The corresponding row address is given by the first and last bit of the input and the column address is given by the middle bit of the input. What is the output of the S-box?

10	0	9	14	6	3	15	5	1	13	12	7	11	4	2	8
13	7	0	9	3	4	6	10	2	8	5	14	12	11	15	1
13	6	4	9	8	15	3	0	11	1	2	12	5	10	14	7
1	10	13	0	6	9	8	7	4	15	14	3	11	5	2	12

Answer

$$\text{Input} = 100\,111_2$$
$$\text{Row address} = (\text{first bit})(\text{last bit}) = 11_2 = 3$$
$$\text{Column address} = \text{middle bits} = 0011_2 = 3$$
So output is given by row three and column 3, that is, 4
So output = 0100

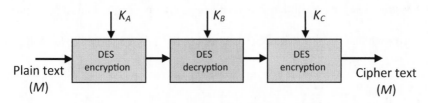

Figure 4.4   Triple DES

There have been many approaches developed to break the DES cipher, yet the best practical attack known is the exhaustive Key search. Concerns over the relatively short Key length of 56 bits led to the DES being strengthened in a simple way that led to the development of Triple DES. The Triple DES ciphers use three iterations, encryption–decryption–encryption, of DES as shown in Figure 4.4. The DES decryption is achieved by using the same 16-round process but with subkeys used in reverse order, K_{16} through K_1.

Apart from DES, there are several other block cipher algorithms in common use. Examples of commonly used algorithms are: Advanced Encryption Standard (AES), Blowfish, Serpent and Twofish [3].

Some applications find a block cipher has drawbacks. For instance, since block ciphers need to have a complete block of data to commence processing, it makes both the encryption and decryption processes slow. A stream cipher is another symmetric key cipher where Plain text bits are typically XOR with a pseudorandom Key stream. Stream cipher can be considered a one-time pad encryption. RC4 is an example of a commonly used stream cipher algorithm.

4.2.2   Public key encryption

Key distribution is an issue for all cryptography. Symmetric key encryption algorithms require a secure initial exchange of secret Keys between sender and receiver and the number of secret Keys required grows with the number of devices in a network. Public Key encryption does not require secure initial exchanges of secret Keys between the sender and receiver.

Public key algorithms involve a pair of Keys called the public Key and the private Key. Each user announces its public Key but retains its private Key confidentially. If user A wishes to send a message to user B, then A encrypts the message using B's public Key. Public Key algorithms are such that it is practically not possible to determine the decryption Key even though the encryption Key is known as it uses one key for encryption and another for decryption. RSA (this acronym stands for Rivest, Shamir and Adleman who first publicly described it) is a widely used public Key algorithm.

The RSA algorithm requires a methodical generation of Keys. The following is the process:

1. Choose two distinct prime numbers p and q.
2. Compute $n = p \times q$.
3. Compute $z = (p - 1)(q - 1)$.

4. Choose an integer e such that it is relatively prime to z (both e and z have no common factors).
5. Find d such that $e \times d \equiv 1 \pmod{z}$ (in other words $e \times d - 1$ should be an integer multiple of z).
6. The pair (e, n) is then released as the public Key and pair (d, n) is used as the private Key.
7. The encryption of message M gives the Cipher text $C = M^e \times Mod(n)$.
8. The decryption is done by computing $M = C^d \times Mod(n)$.

Example: Let's consider that user B wishes to transmit character 'D' to user A. The following is the steps involved in the RSA algorithm:

1. A selects p and q. Assume that he chooses $p = 11$ and $q = 3$, p and q are prime numbers.
2. A computes $n = p \times q = 33$.
3. A computes $z = (p - 1) \times (q - 1) = 20$.
4. A chooses $e = 7$ which is relatively prime to $z = 20$.
5. A finds $d = 3$ that makes $3 \times 7 = 21 \equiv 1 \pmod{20}$.
6. A publishes pair $(7, 33)$ as the public Key.
7. B uses A's public Key $(33, 7)$ and computes $C = 4^7$ Mod $33 = 16$ and transmits 16 to A (D is the 4th character in the alphabet).
8. A computes $M = 16^3$ Mod $33 = 4$ (A obtains D).

4.3 Authentication

Authentication is required to verify the identities of communicating parties to avoid imposters gaining access to information. When user A receives a communication from user B, A needs to verify that it is actually B, but not someone else masquerading as B, who is talking to him. Detailed descriptions of authentication methods can be found in [1].

4.3.1 Authentication based on shared secret key

Assume that A and B wish to establish a communication session. Prior to exchanging data, they need to authenticate each other. The steps involved in this method are shown in Figure 4.5.

a) A indicates to B that it wishes to communicate with B and sends his identity with a large random number (N_A) in Plain text.
b) B encrypts N_A using a secret Key known to A and B and sends Cipher text ($E_{KAB}(N_A)$) together with another large random number (N_B) to A as Plain text.
c) A decrypts the Cipher text received to check whether it gets the same number (N_A) that he sent to B and encrypts the number N_B using a shared secret Key and sends the Cipher text to B.
d) B decrypts the received Cipher text using a shared secret Key and checks whether he gets the same number (N_B) as that he sent.

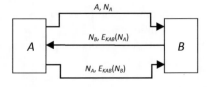

Figure 4.5 Authentication based on secret shared key [1]

a) *A* indicates to KDC that it wishes to communicate with *B* using a session Key K_s and sends his identity with *B*'s identity and K_s encrypted using the shared secret Key *KA* of *A* and KDC (E_{KA} (B, K_s)).

b) KDC decrypts the Cipher text received from *A* and extract proposed session Key K_S and sends it together with *A*'s identity after encrypting them using the shared secret Key *KB* of *B* and KDC (E_{KB} (A, K_s)).

c) *B* decrypts the Cipher text received from KDC and extracts *A*'s identity and session Key K_s and uses it for the communication with *A*.

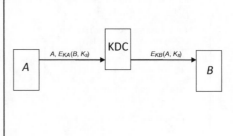

Figure 4.6   Authentication based on Key Distribution Centre key [1]

4.3.2   Authentication based on key distribution centre

This method involves a trusted key distribution centre (KDC) that supports the authentication. KDC and each of its users have a shared secret Key that is used to communicate between them. The steps involved are shown in Figure 4.6.

4.4   Digital signatures

A digital signature allows the signing of digital messages by the Sender in such a way that:

1. The Receiver can verify the claimed identity of the Sender (authentication).
2. The Receiver can prove and the Sender cannot deny that the message has been sent by the specific user (non-repudiation).
3. The Receiver cannot modify the message and claim that the modified message is the one that was received from the Sender.

Approaches for digitally signing messages include the secret Key signature, the public Key signature and the message digest.

4.4.1   Secret key signature

In this approach a Central Authority (CA) that is trusted by users is used to sign messages. The operation of this approach is explained in Figure 4.7. It is possible for someone to capture the encrypted message going from CA to *B* and replay it. To avoid such attacks, a time stamp and message number can be included.

4.4.2   Public key signature

This approach uses the property of a public Key algorithm that makes: $E_{KEB}(E_{KDA}(M)) = E_{KDB}(E_{KEA}(M)) = M$. This property is used as shown in Figure 4.8.

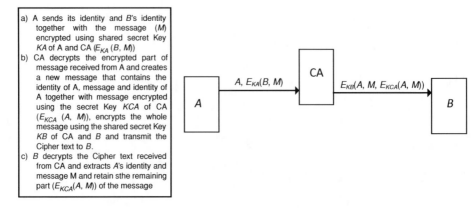

a) A sends its identity and B's identity together with the message (M) encrypted using shared secret Key KA of A and CA (E_{KA} (B, M))
b) CA decrypts the encrypted part of message received from A and creates a new message that contains the identity of A, message and identity of A together with message encrypted using the secret Key KCA of CA (E_{KCA} (A, M)), encrypts the whole message using the shared secret Key KB of CA and B and transmit the Cipher text to B.
c) B decrypts the Cipher text received from CA and extracts A's identity and message M and retain sthe remaining part (E_{KCA}(A, M)) of the message

Figure 4.7   Signing a message with a secret Key [1]

4.4.3   Message digest

Both the methods discussed previously sign the whole document. This is inefficient if the messages are long because retaining proof consumes a large space. Signing the message digest, a miniature version of the message, is more efficient.

The Message Digest (MD) is created using a hash function which has the following properties:

1. It is a one-way function (that is, from the message the digest can be created easily but the reverse is not practically possible).
2. It practically provides one-to-one mapping to make sure that the creation of two messages that produce the same digest is impossible.

The approach of signing with the message digest is explained in Figure 4.9. Commonly used hash functions are MD5 (Message Digest 5) which produces a 120-bit digest and SHA (Secure Hash Algorithm) which produces a 160-bit digest.

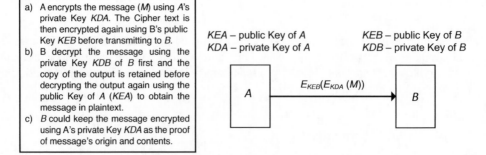

a)   A encrypts the message (M) using A's private Key KDA. The Cipher text is then encrypted again using B's public Key KEB before transmitting to B.
b)   B decrypt the message using the private Key KDB of B first and the copy of the output is retained before decrypting the output again using the public Key of A (KEA) to obtain the message in plaintext.
c)   B could keep the message encrypted using A's private Key KDA as the proof of message's origin and contents.

KEA – public Key of A   KEB – public Key of B
KDA – private Key of A   KDB – private Key of B

Figure 4.8   Signing a message with a public Key [1]

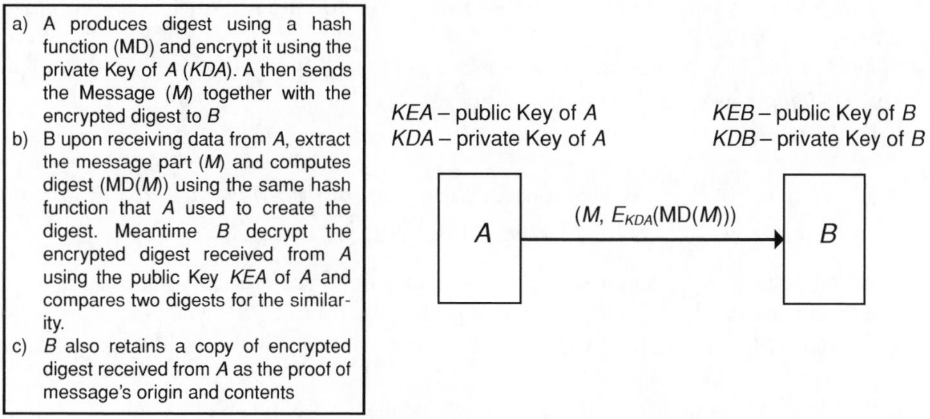

a) A produces digest using a hash function (MD) and encrypt it using the private Key of A (*KDA*). A then sends the Message (*M*) together with the encrypted digest to B

b) B upon receiving data from A, extract the message part (*M*) and computes digest (MD(*M*)) using the same hash function that A used to create the digest. Meantime B decrypt the encrypted digest received from A using the public Key *KEA* of A and compares two digests for the similarity.

c) B also retains a copy of encrypted digest received from A as the proof of message's origin and contents

KEA – public Key of *A*   *KEB* – public Key of *B*
KDA – private Key of *A*   *KDB* – private Key of *B*

$$A \xrightarrow{\ (M, E_{KDA}(MD(M)))\ } B$$

Figure 4.9   Signing a message with the message digest [1]

4.5   Cyber security standards

There are several standards which apply to the security of substation equipment and many are under development. For overall security assessment, the standard ISO 27001 is widely used and specifies the assessment of risks for a system of any sort and the strategy for developing the security system to mitigate those risks. Furthermore, ISO 28000 specifies security management specifically for a supply chain system.

4.5.1   IEEE 1686: IEEE Standard for substation intelligent electronic devices (IEDs) cyber security capabilities

This standard originated from an IED security effort of the NERC CIP (North America Electric Reliability Corporation – Critical Infrastructure Protection). The standard is applicable to any IED where the user requires "*security, accountability, and auditability in the configuration and maintenance of the IED*".

The standard proposes different mechanisms to protect IEDs. The IED shall:

• be protected by unique user ID and password combinations. The password should be a minimum of 8 characters with at least one upper and lower cases, one number and one alpha-numeric character.
• not have any means to defeat or circumvent the user created ID/password. The mechanisms such as "*embedded master password, chip-embedded diagnostic routines that automatically run in the event of hardware or software failures, hardware bypass of passwords such as jumpers and switch settings*" shall not be present.
• support different level of utilisation of IED functions and features based on individual user-created ID/password combinations.
• "*have a time-out feature that automatically logs out a user.*

- *record in a sequential circular buffer (first in, first out) an audit trail listing events in the order in which they occur.*
- *monitor security-related activity and make the information available through a real-time communication protocol for transmission to SCADA."*

4.5.2   IEC 62351: Power systems management and associated information exchange – data and communications security

IEC 62351 is a series of documents which specifies the types of security measures for communication networks and systems including profiles such as TCP/IP, Manufacturing Message Specification (MMS) and IEC 61850. Some security measures included in the standard are:

- authentication to minimise the threat of attacks, some types of bypassing control, carelessness and disgruntled employee actions;
- authentication of entities through digital signatures;
- confidentiality of authentication keys and messages via encryption;
- tamper detection;
- prevention of playback and spoofing;
- monitoring of the communications infrastructure itself.

References

[1] Tanenbaum, A.S. (2001) *Computer Networks*, Prentice-Hall of India, New Delhi.
[2] Elbirt, A.J. (2009) *Understanding and Applying Cryptography and Data Security*, CRC Press, New York.
[3] Nadeem, A. and Javed, M.Y. (2003) *A performance comparison of data encryption algorithms*. First International Conference on Information and Communication Technologies, December 2003, Cairo, Egypt.

Part II

Sensing, Measurement, Control and Automation Technologies

5

Smart Metering and Demand-Side Integration

5.1 Introduction

In many countries, the power infrastructure is ageing and is being increasingly heavily used as demand for electricity rises. This overloading will worsen as large numbers of electric vehicles, heat pumps and other new loads use low-carbon energy from the electric power system. Obtaining planning permission for the installation of new power system equipment, particularly overhead lines, is becoming increasingly difficult. Therefore, demand-side programmes have been introduced widely to make better use of the existing power supply infrastructure and to control the growth of demand.

The dual aims of reducing CO_2 emissions and improving energy security (energy policy goals in many countries) coincide in the increasing use of renewable energy for electricity generation. However, connection of a large amount of intermittent renewable generation alters the pattern of the output of central generation and the power flows in both transmission and distribution circuits. One solution to this increase in variability is to add large-scale energy storage devices to the power system. This is often not practical at present due to technical limitations and cost. Therefore, flexibility in the demand side is seen as another way to enable the integration of a large amount of renewable energy.

Load control or load management has been widespread in power system operation for a long time with a variety of terminology used to describe it. The name Demand-Side Management (DSM) has been used since the 1970s for a systematic way of managing loads [1]. Later on, Demand Response (DR), Demand-Side Response (DSR), Demand-Side Bidding (DSB) and Demand Bidding (DB) were used to describe a range of different demand side initiatives (see Section 5.7 for definitions). To avoid the confusion caused by such overlapping concepts and terminologies, as recommended by CIGRE, Demand-Side Integration (DSI) is used in this chapter to refer to all aspects of the relationships between the electric power system, the energy supply and the end-user load.

Effective implementation of DSI needs an advanced ICT (Information and Communication Technology) infrastructure and good knowledge of system loads. However, the electro-mechanical meters that are presently installed in domestic premises have little or no

Smart Grid: Technology and Applications, First Edition.
Janaka Ekanayake, Kithsiri Liyanage, Jianzhong Wu, Akihiko Yokoyama and Nick Jenkins.
© 2012 John Wiley & Sons, Ltd. Published 2012 by John Wiley & Sons, Ltd.

communication ability and do not transmit information of the load in real time. Smart metering refers to systems that measure, collect, analyse, and manage energy use using advanced ICT. The concept includes two-way communication networks between smart meters and various actors in the energy supply system.[1] The smart meter is seen to facilitate DSI through providing real-time or near-real-time information exchange and advanced control capabilities.

5.2 Smart metering

5.2.1 Evolution of electricity metering

Electricity meters are used to measure the quantity of electricity supplied to customers as well as to calculate energy and transportation charges for electricity retailers and network operators. The most common type of meter is an accumulation meter, which records energy consumption over time. Accumulation meters in consumer premises are read manually to assess how much energy has been used within a billing period. In recent years, industrial and commercial consumers with large loads have increasingly been using more advanced meters, for example, interval meters which record energy use over short intervals, typically every half hour. This allows the energy suppliers to design tariffs and charging structures that reflect wholesale prices and helps the customers understand and manage their pattern of electricity demand. Smart meters are even more sophisticated as they have two-way communications and provide a real-time display of energy use and pricing information, dynamic tariffs and facilitate the automatic control of electrical appliances. Figure 5.1 shows the evolution of electrical metering, from simple electro-mechanical accumulation metering to advanced smart metering.

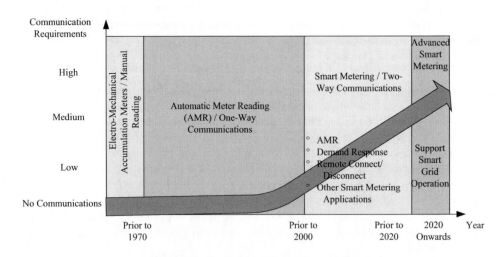

Figure 5.1   Evolution of electricity metering

---

[1] These actors can include energy suppliers, network operators, other emerging service companies, for example, aggregators, and also intelligent home area energy management systems connected on the premises side of a smart meter.

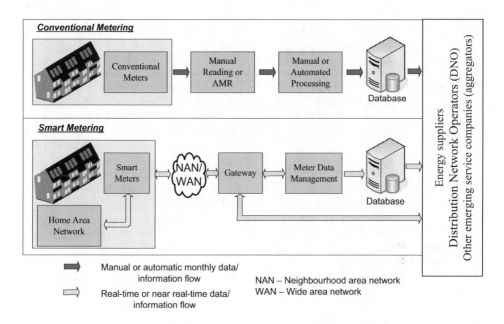

Figure 5.2   Conventional and smart metering compared

It can be seen from Figure 5.1 that manual reading was widespread prior to the year 2000. A number of Automatic Meter Reading (AMR) programs were developed around this time where energy consumption information was transmitted monthly from the meters to the energy supplier and/or network operator using low-speed one-way communications networks. Since 2000, there has been a dramatic increase in the performance of the metering infrastructure being installed. One-way communication of meter energy use data, AMR, has given way to more advanced two-way communications supporting applications such as varying tariffs, demand-side bidding and remote connect/disconnect. The Smart Grid vision represents a logical extension of these capabilities to encompass two-way broadband communications supporting a wide range of Smart Grid applications including distribution automation and control as well as power quality monitoring.

The differences between conventional metering and smart metering are shown schematically in Figure 5.2. Smart meters have two-way communications to a Gateway and/or a Home Area Network (HAN) controller. The Gateway[2] allows the transfer of smart meter data to energy suppliers, Distribution Network Operators (DNOs) and other emerging energy service companies. They may receive meter data through a data management company or from smart meters directly.

The benefits of advanced metering are listed in Table 5.1. Short-term benefits, particularly for the energy suppliers and metering operators, can be obtained from AMR and Automatic Meter Management (AMM). Longer-term benefits arise from the additional functions of smart metering that lead on to the use of smart meters in the Smart Grid.

---

[2] The Gateway provides the bridge between the Smart Meters, the Meter Data Management system and other actors.

Table 5.1   Benefits of advanced metering

	Energy suppliers and network operator benefits	All benefit	Customer benefits
Short-term	Lower metering costs and more frequent and accurate readings	Better customer service Variable pricing schemes	Energy savings as a result of improved information
	Limiting commercial losses due to easier detection of fraud and theft	Facilitating integration of DG and flexible loads	More frequent and accurate billing
Longer-term	Reducing peak demand via DSI programs and so reducing cost of purchasing wholesale electricity at peak time	More reliable energy supply and reduced customer complaints	Simplification of payments for DG output
	Better planning of generation, network and maintenance	Using ICT infrastructure to remotely control DG, reward consumers and lower costs for utility	Additional payments for wider system benefits
	Supporting real-time system operation down to distribution levels	Facilitating adoption of electric vehicles and heat pumps, while minimising increase in peak demand	Facilitating adoption of home area automation for more comfortable life while minimising energy cost
	Capability to sell other services (e.g. broadband and video communications)		

5.2.2   Key components of smart metering

Smart metering consists of four main components: smart meters, a two-way communication network, a Meter Data Management system (MDM), and HAN. In order to integrate smart metering into the operation and management of the power system, interfaces to a number of existing systems are required, for example, the interface to the load forecasting system, the Outage Management System (OMS), and a Customer Information System (CIS) (see Chapter 7 for more details).

5.3   Smart meters: An overview of the hardware used

A traditional electro-mechanical meter has a spinning aluminium disc and a mechanical counter display that counts the revolutions of the disc. The disc is situated in between two coils, one fed with the voltage and the other fed with the current of the load. The current coil produces a magnetic field, ϕ_I and the voltage coil produces a magnetic field, ϕ_V. The forces acting on the disc due to the interaction between the eddy currents induced by ϕ_I and the magnetic field ϕ_V and the eddy currents induced by ϕ_V and the magnetic field ϕ_I produce a torque. The torque is proportional to the product of instantaneous current and voltage, thus to the power. The

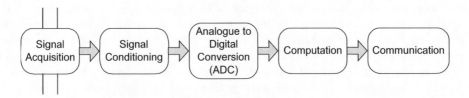

Figure 5.3   Functional block diagram of a smart meter

number of rotations of the disc is recorded on the mechanical counting device that gives the energy consumption.

The replacement of electro-mechanical meters with electronic meters offers several benefits. Electronic meters not only can measure instantaneous power and the amount of energy consumed over time but also other parameters such as power factor, reactive power, voltage and frequency, with high accuracy. Data can be measured and stored at specific intervals. Moreover, electronic meters are not sensitive to external magnets or orientation of the meter itself, so they are more tamperproof and more reliable.

Early electronic meters had a display to show energy consumption but were read manually for billing purposes. More recently electronic meters with two-way communications have been introduced. Figure 5.3 provides a general functional block diagram of a smart meter. In Figure 5.3, the smart meter architecture has been split into five sections: signal acquisition, signal conditioning, Analogue to Digital Conversion (ADC), computation and communication.

5.3.1   Signal acquisition

A core function of the smart meter is to acquire system parameters accurately and continuously for subsequent computation and communication. The fundamental electrical parameters required are the magnitude and frequency of the voltage and the magnitude and phase displacement (relative to the voltage) of current. Other parameters such as the power factor, the active/reactive power, and Total Harmonic Distortion (THD) are computed using these fundamental quantities.

Current and voltage sensors measure the current into the premises (load) and the voltage at the point of supply. In low-cost meters the measuring circuits are connected directly to the power lines, typically using a current-sensing shunt resistor on the current input channel and a resistive voltage divider on the voltage input channel (Figure 5.4).

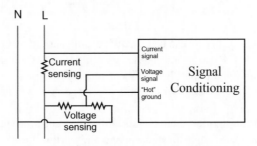

Figure 5.4   Current and voltage sensing

The current sensing shunt is a simple high stability resistor (typically with resistance between $100\ \mu\Omega$ and $500\ m\Omega$) with the voltage drop across it proportional to the current flowing through it. The current rating of this shunt resistor is limited by its self-heating so it is usually used only in residential meters (maximum current less than 100 A). In order to match the voltage across the current sensing resistor (which is very small) with the Analogue to Digital Converter (ADC), a Programmable Gain Amplifier (PGA) is used in the signal conditioning stage before the ADC (normally integrated within a single chip with the ADC).

The voltage resistive divider gives the voltage between the phase conductor and neutral. The alloy Manganin is suitable for the resistive divider due to its near constant impedance over typical operating temperature ranges.

Example 5.1

A specification sheet of a smart meter states that its rated current is 100 A and power dissipation is 3 W. It employs a current-sensing resistor of $200\ \mu\Omega$. When the load current is at the rated value of the meter, calculate:

1. the power dissipation in all the other components of the meter;
2. the voltage across the current-sensing resistor;
3. the gain of the PGA to match with an ADC having a full scale of 5 V.

Answer

1. The power dissipated in the current-sensing resistor is given by:

$$P_R = I^2R = (100)^2 \times 200 \times 10^{-6} = 2\text{ W}$$

Therefore, the power consumed by other components (the microcontroller, display, and so on) is:

$$P_{remain} = (3 - 2)\text{W} = 1\text{ W}$$

2. Voltage across the current-sensing resistor at full-load current is:

$$I \times R = 100 \times 200 \times 10^{-6} = 0.02\text{ V}$$

3. Gain of the PGA $= 5/0.02 = 250$.

A Current Transformer (CT) can also be used for sensing current and providing isolation from the primary circuit. A CT can handle higher currents than a shunt and also consumes less power. The disadvantages are that the nonlinear phase response of the CT can cause power or energy measurement errors at low currents and large power factors, and also the higher meter cost.

Some applications may require smart meters with high precision over a wide operating range. For such applications more sophisticated voltage and current measuring techniques such as

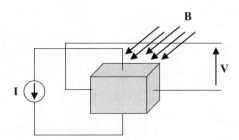

Figure 5.5   Simplified diagram of a Hall Effect sensor [2]

Rogowski coils, optical methods and Hall Effect sensors may be used. Detailed explanations of the Rogowski coil and optical methods are given in Chapter 6.

The Hall Effect is a phenomenon in which a magnetic field across a thin conductive material, with a known current flowing (I), causes a voltage (V) across the material, proportional to the flux density (B), as shown in Figure 5.5. This voltage is measured perpendicular to the direction of current flow. Hall Effect sensors can be used to measure the magnetic field around a conductor and therefore the current flowing within it.

5.3.2   Signal conditioning

The signal conditioning stage involves the preparation of the input signals for the next step in the process, ADC. The signal conditioning stage may include addition/subtraction, attenuation/amplification and filtering. When it comes to physical implementation, the signal conditioning stages can be realised as discrete elements or combined with the ADC as part of an Integrated Circuit. Alternatively the stages can be built into a 'System on a Chip' architecture with a number of other functions.

In many circumstances the input signal will require attenuation, amplification or the addition/subtraction of an offset such that its maximum magnitude lies within the limits of the inputs for the ADC stage.

To avoid inaccuracy due to aliasing, it is necessary to remove components of the input signal above the Nyquist frequency (that is, half the sampling rate of the ADC). Therefore, prior to input to the ADC stage, a low pass filter is applied to the signal. The sampling frequency is determined by the functions of the meter. If the meter provides fundamental frequency measurements (currents, voltage and power) and in addition harmonic measurements, then the sampling frequency should be selected sufficiently high so as to obtain harmonic components accurately.

Example 5.2

A smart meter displays current harmonic measurements up to the 5th harmonic component. What should be the minimum sampling frequency used in the signal conditioning stage? Assume that the frequency of the supply is 50 Hz.

Answer

The frequency of the 5th harmonic component = 5 × 50 Hz = 250 Hz. In order to capture up to 5th harmonic component, the signal should be filtered by an anti-aliasing filter [3] with a cut-off frequency of 250 Hz.

According to the Nyquist criteria, the minimum sampling frequency should then be at least = 2 × 250 Hz = 500 Hz. This is shown in Figure 5.6, where f_s is the sampling frequency.

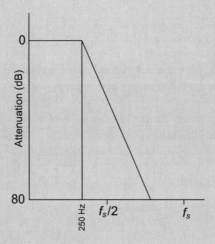

Figure 5.6   Characteristic of an anti-aliasing filter

5.3.3   *Analogue to digital conversion*

Current and voltage signals obtained from the sensors are first sampled and then digitised to be processed by the metering software. Since there are two signals (current and voltage) in a single phase meter, if a single ADC is used, a multiplexer is required to send the signals in turn to the ADC.

The ADC converts analogue signals coming from the sensors into a digital form. As the number of levels available for analogue to digital conversion is limited, the ADC conversion always appears in discrete form. Figure 5.7 shows an example of how samples of a signal are digitised by a 3-bit ADC. Even though 3-bit ADCs are not available, here a 3-bit ADC was used to illustrate the operation of an ADC simply. The 3-bit ADC uses 2^3 (= 8) levels thus any voltage between −0.8 and −0.6 V is represented by 000 (the most negative range is assigned 000). In other word, −0.8, −0.75, −0.7, and −0.65 are all represented by 000. Similarly, voltage between −0.6 and −0.4 V is represented by 001, and so on.

The resolution of an ADC is defined as:

$$\text{Resolution} = \text{Voltage range}/2^n; \text{ where } n \text{ is the number of bits in the ADC.}$$

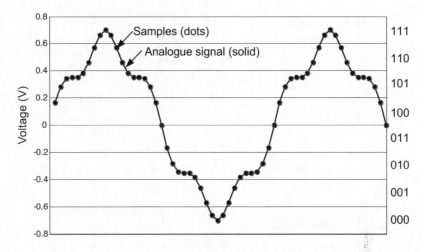

Figure 5.7   Operation of a 3-bit ADC

For the 3-bit ADC shown in Figure 5.7, the voltage range is 1.6 V (-0.8 to 0.8) and therefore the resolution is $1.6/2^3 = 0.2$ V. The higher the number of bits used in the ADC, the lower the resolution. For example, if an 8-bit ADC (typically 8-, 16- and 32-bit ADCs are available) is used, the resolution is $1.6/2^8 = 6.25$ mV.

Example 5.3

A smart meter uses the same 16-bit analogue to digital converter for both current and voltage measurements. It uses a 100 : 5 A CT for current measurements and 415: 10 V potential divider for voltage measurements. When the meter shows a current measurement of 50 A and a voltage measurement of 400 V, what is the maximum possible error in the apparent power reading due to the quantisation of the voltage and current signals?

Answer

Current range to the ADC: 0–5 A
Resolution of the ADC: $5/2^{16} = 76$ μA
So the maximum quantisation error of the current is 76 μA.
50 A passes through the primary of the CT and so the ADC reads: $50 \times 5/100 = 2.5$ A
Voltage range to the ADC: 0–10 V
Resolution of the ADC: $10/2^{16} = 152$ μV
So the maximum quantisation error of the voltage is 152 μV.
400 V is read by voltage divider and so the ADC reads; $400 \times 10/415 = 9.64$ V

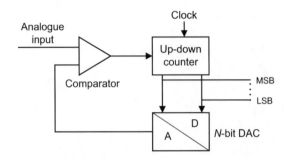

Figure 5.8   Simplified successive approximation ADC architecture [3]

The apparent power reading is:

$$((V + \Delta V)(I + \Delta I) = VI + V\Delta I + I\Delta V + \Delta V\Delta I) \approx VI + V\Delta I + I\Delta V$$

Therefore, the maximum possible error in the apparent power reading due to the quantisation is:

$$V\Delta I + I\Delta V = 9.64 \times 76 \times 10^{-6} + 2.5 \times 152 \times 10^{-6} = 1.11 \text{ mVA}.$$

There are many established methods for conversion of an analogue input signal to a digital output [3, 4, 5]. The majority of the methods involve an arrangement of comparators and registers with a synchronising clock impulse. The most common ADCs for metering use the successive approximation and the sigma-delta method.

5.3.3.1   The successive approximation method

In a successive approximation ADC, shown in Figure 5.8, the up-down counter initially sets the Most Significant Bit (MSB) of its output to 1 while keeping all other outputs at zero. The counter output is converted into an analogue signal using a Digital to Analogue Converter (DAC) and compared with the analogue input by a comparator. If the analogue input signal is larger than the DAC output, then the up-down counter sets the MSB and the next bit to 1 and the comparison is repeated. If the analogue signal is smaller than the DAC output, then the MSB is reset to zero and the next bit is set to 1. This process is repeated until the analogue input signal is the same as the DAC output. At that point the DAC input will be same as the digitised value of the analogue signal.

5.3.3.2   The sigma-delta method

The sigma-delta converter consists of an integrator, a latched comparator, and a single-bit DAC, as shown in Figure 5.9. The output of the DAC (signal at E) is subtracted from the input signal, A. The resulting signal, B, is then integrated, and the integrator output voltage (signal

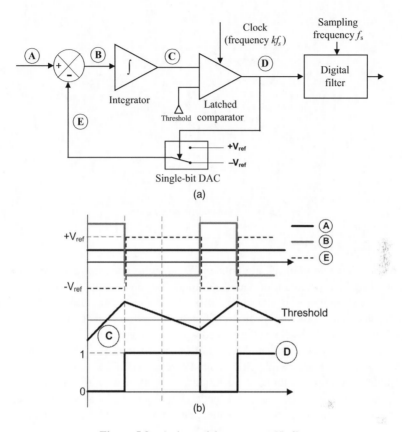

Figure 5.9 A sigma-delta converter [5, 6]

at C) is converted to a single-bit digital output (1 or 0) by the comparator (signal at D). The high frequency bit stream at D (at frequency kf_s where $k > 1$) is finally divided by the digital filter thus giving a series of bits corresponding to the digitised value of the analogue signal at every sampling frequency (f_s).

In order to explain the operation of this circuit, assume that the signal at A is $V_{ref}/2$, that the output at E is $-V_{ref}$ with signal at C below the threshold (corresponding to logic '0'). Initially the output of the summation (B) is $3V_{ref}/2$ ($V_{ref}/2+V_{ref}$), and the output of the integrator increases linearly. At the first clock pulse, as the integrator output is greater than its threshold, it gives logic '1' (see Figure 5.9b). Since the corresponding DAC output is $+V_{ref}$, the signal at B now becomes negative ($V_{ref}/2-V_{ref}$). Therefore, the integrator output reduces linearly (rate determined by the difference of signals at A (actual signal) and E (analogue value of the digitised signal)). At the next clock the output of the integrator is greater than the threshold, the output remains at '1'. This negative feedback loop works such that the signal at D becomes the digitised signal of input signal at A.

In real implementations instead of the integrator in Figure 5.9a, a digital filter is employed. Further, the comparator is also in digital form, thus analogue errors are not accumulated [6].

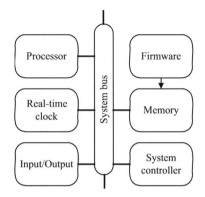

Figure 5.10   Computation overview block diagram

5.3.4   Computation

The computation requirements are split into arithmetic operations on input signals, time-stamping of data, preparation of data for communication or output peripherals, handling of routines associated with irregular input (such as payment, tamper detection), storage of data, system updates and co-ordinating different functions. The block diagram shown in Figure 5.10 shows different functional blocks associated with the computation functions of a smart meter.

Due to the relatively large number of arithmetic operations (Table 5.2) required for the derivation of the parameters, a Digital Signal Processor (DSP) is used.

In addition to routine arithmetic operations, a meter deals with a large number of other procedures (that is, payment, tamper detection, system updates, user interactions) as well as other routine tasks (for example, the communication of billing information). Therefore, a high degree of parallelism (the ability to perform multiple tasks, involving the same data sets, simultaneously) and/or buffering (the ability to temporarily pause arithmetical operations so that other needs can be attended to) is required.

Table 5.2   Arithmetic operation required for different parameters

Required parameter	Operation type
Instantaneous voltage	Multiplication
Instantaneous current	Multiplication
Peak voltage/current	Comparison
System frequency	Zero detection, Fourier analysis
RMS voltage/current	Multiplication
Phase displacement	Zero detection, comparison
Power factor	Trigonometric function
Instantaneous apparent power	Multiplication
Instantaneous real power	Multiplication
Instantaneous reactive power	Multiplication
Energy use/production	Integration
Harmonic voltage distortion	Fourier analysis
Total harmonic distortion	Multiplication and addition

For computation, volatile memory (where information is lost on loss of power supply) and non-volatile [3] memory is needed. Volatile memory is used for temporary storage of data to support the processor(s) as operations are undertaken. The amount of volatile memory used depends on the quantity, rate and complexity of computation and the rate of communication to/from ports. A certain amount of non-volatile memory is typically required to store specific information, such as the unit serial number and maintenance access key codes. Additionally data related to energy consumption should be retained until successful communication to the billing company has been achieved.

In order that the acquired data can be meaningfully interrogated, a time reference must be appended to each sample and/or calculated parameter. For this purpose a real-time clock is used. The accuracy of the real-time clock can vary with temperature. In order to maintain this function during system power losses or maintenance, a dedicated clock battery is typically used.

5.3.5   Input/Output

A smart meter has a display that presents information in the form of text and graphs for the human user. Liquid Crystal Displays (LCD) and the Light Emitting Diodes (LED) are preferred for their low cost and low power consumption requirements. Both display types are available in seven-segment, alphanumeric and matrix format. LEDs are relatively efficient light sources, as they produce a significant amount of light when directly polarised (at relatively low voltages: 1.2–1.6 V), and a current of a few milliamps is applied.

Smart meters provide a small key pad or touch screen for human–machine interaction, for instance, to change the settings of a smart meter so as to select the smart appliance to be controlled or to select payment options.

As smart meters require calibration due to variations in voltage references, sensor tolerances or other system gain errors, a calibration input is also provided. Some meters also provide remote calibration and control capability through communication links.

Energy consumption and tariffs may be displayed on a separate customer display unit located in an easily visible location within the residence (for example, the kitchen). This is to encourage customers to reduce their energy use, either throughout the year or at times of peak demand when generation is short. Research is ongoing to determine the most effective way to display information to encourage customers to take notice of their energy consumption, and/or the signals from the suppliers to restrict demand at times of generation shortage. Approaches that have been used include displays using:

- three coloured lights (resembling traffic lights) or a globe that changes colour to signal changes in tariffs. This is used with a Time of Use Tariff to control peak demand.
- a digital read-out or analogue display resembling a car speedometer showing energy use;
- a continuously updated chart showing energy use and comparison with a previous period, for example, yesterday or last week.

It is hoped that customers will manage and reduce their energy consumption when they are provided with more accurate, up-to-date information, also that any reduction made soon after the display is installed will be maintained. Trials indicate that initial reductions in electrical energy use of up to 10 per cent may be possible but maintaining this level of reduction requires careful design of the displays and tariffs but also other interventions such as outreach programmes to customers that provide advice on how to reduce energy consumption.

5.3.6   Communication

Smart meters employ a wide range of network adapters for communication purposes. The wired options include the Public Switched Telephone Network (PSTN), power line carrier, cable modems and Ethernet. The wireless options include ZigBee, infrared, and GSM/GPRS/CDMA Cellular. These techniques are described in Chapter 2.

5.4   Communications infrastructure and protocols for smart metering

A typical communications architecture for smart metering is shown in Figure 5.11. It has three communications interfaces: Wide Area Network (WAN), Neighbourhood Area Network (NAN) and Home Area Network (HAN).

5.4.1   Home-area network

A Home-Area Network (HAN) is an integrated system of smart meter, in-home display, micro-generation, smart appliances, smart sockets, HVAC (Heating, Ventilation, Air Conditioning) facilities and plug-in hybrid/electric vehicles. A HAN uses wired or wireless communications and networking protocols to ensure the interoperability of networked appliances and the interface to a smart meter. It also includes security mechanisms to protect consumer data and the metering system.

A HAN enables centralized energy management and services as well as providing different facilities for the convenience and comfort of the household. Energy management functions provided by HAN include energy monitoring and display, controlling the HVAC system and controlling smart appliances and smart plugs. The services provided by HAN for the convenience of the household can include scheduling and remote operation of household

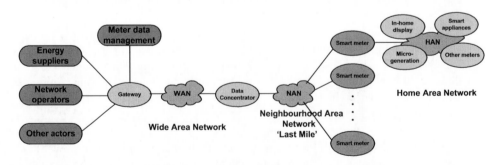

Figure 5.11   Smart metering communications

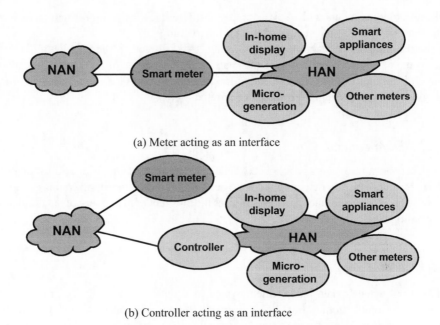

(a) Meter acting as an interface

(b) Controller acting as an interface

Figure 5.12   Interface between the HAN and NAN

appliances as well as household security systems. Home-based multimedia applications such as media centres for listening to music, viewing television and movies require broadband Internet access across the HAN. A separate HAN used for energy services can coexist with the broadband Internet system but there is some expectation that the systems will be merged in the future.

It is expected that HAN will provide benefits to the utilities through demand response and management and the management of micro-generation and the charging of electric vehicles. In order to provide demand management functions and demand response, two options are being actively considered in different countries (Figure 5.12). One option is to use the smart meter as the interface to the suppliers, network operators and other actors. The other option is to use a separate control box [7, 8] which is directly interfaced to the outside world through the NAN and WAN.

5.4.2   Neighbourhood area network

The primary function of the Neighbourhood Area Network (NAN) is to transfer consumption readings from smart meters. The NAN should also facilitate diagnostic messages, firmware upgrades and real-time or near real-time messages for the power system support. It is anticipated that the data volume transferred from a household for simple metering is less than 100 kB[3] per day and firmware upgrades may require 400 kB of data to be transferred [9].

---

[3] 1 kB = 1000 bytes.

However, these numbers will escalate rapidly if different real-time or near real-time smart grid functions are added to the smart metering infrastructure.

The communication technology used for the NAN is based on the volume of data transfer. For example, if ZigBee technology which has a data transfer rate of 250 kb/s is used, then each household would use the communication link only a fraction of a second per day to transfer energy consumption data to the data concentrator.

5.4.3 Data concentrator

The data concentrator acts as a relay between the smart meters and the gateway. It manages the meters by automatically detecting them, creates and optimises repeating chains (if required to establish reliable communication), coordinates the bi-directional delivery of data, and monitors the conditions of the meters.

5.4.4 Meter data management system

The core of a meter data management system is a database. It typically provides services such as data acquisition, validation, adjustment, storage and calculation (for example, data aggregation), in order to provide refined information for customer service and system operation purposes such as billing, demand forecasting and demand response.

A major issue in the design and implementation of a meter data management system is how to make it open and flexible enough to integrate to existing business/enterprise applications and deliver better services and more value to customers while ensuring data security. Besides the common database functionalities, a meter data management system for smart metering also provides functions such as remote meter connection/disconnection, power status verification, supply restoration verification and on-demand reading of remote smart meters.

5.4.5 Protocols for communications

Currently various kinds of communication and protocol types are used for smart metering. For example, a combination of Power Line Carrier (PLC) and GPRS communication is used in Denmark, Finland and Italy [10]. In these European examples, PLC is used between the meter and data concentrator as the last mile technology and GPRS is used between the concentrator and gateway to the data management system.

Table 5.3 summarizes the characteristics of the most commonly used protocols for demand-side applications, including local AMR, remote AMR, smart metering and home area automation. In Table 5.3, 'Y' means applicable, and a blank means not applicable or the information is still not available. With local AMR, the meter readings are collected by staff using hand-held devices and with remote AMR the meter readings are collected from a distance through communication links. For most protocols listed in Table 5.3, the data frame size is also shown.

The important factors for consideration when assessing communication protocols for smart metering are summarised in Table 5.4.

Table 5.3   Main smart metering protocols

Protocol	Local AMR	Remote AMR	Smart metering	HAN	Estimated frame size (bytes)
TCP/IP		Y	Y	Y	50
IEC 62056	Y	Y	Y	Y	14
SML	Y	Y	Y	Y	14
IEC 61334 PLC		Y	Y		45
EN 13757 M-Bus	Y	Y	Y	Y	27
SITRED	Y	Y	Y		45
PRIME	Y	Y	Y		8
Zigbee Smart Energy			Y	Y	25
EverBlu	Y	Y	Y		
OPERA/UPA		Y	Y		24
IEC 62056-21 'FLAG'	Y	Y			22
IEC 62056-21 'Euridis'	Y	Y			45
ANSI C12.22		Y	Y		64

Table 5.4   Important factors for assessment of smart metering communication protocols [11]

Criteria	Description
Openness	Availability of protocol specifications. Status of controlling body
Interoperability	Extent of ability to interact with other standards, applications and protocols
Scalability/Adaptability	Ease with which the protocol can be extended or changed
Intended function	The intended function of the smart meter such as communicating data to a central entity for billing or communicating data to a third party for other market operations
Maturity	The stage at which the protocol is in its development
Performance	The speed and efficiency with which the protocols operate
Security	Existence of known security vulnerabilities

Detailed communication techniques used in smart metering are introduced in Chapters 2 and 3.

5.5   Demand-side integration

Demand-Side Integration (DSI) is a set of measures to use loads and local generation to support network operation/management and improve the quality of power supply. DSI can help defer investment in new infrastructure by reducing system peak demand. In practice, the potential of DSI depends on: availability and timing of information provided to consumers, the duration and timing of their demand response, performance of the ICT infrastructure, metering, automation of end-use equipment and pricing/contracts.

There are various terms in use in the demand side, whose meanings are closely related to each other but with slightly different focuses. Some widely used definitions are:

- *Demand-Side Management (DSM)*: utility activities that influence customer use of electricity. This encompasses the planning, implementation and monitoring of activities designed to encourage consumers to change their electricity usage patterns.
- *Demand Response (DR)*: mechanisms to manage the demand in response to supply conditions.
- *Demand-Side Participation*: a set of strategies used in a competitive electricity market by end-use customers to contribute to economic, system security and environmental benefits.

In this book, the overall technical area of the efficient and effective use of electricity in support of the power system and customer needs is discussed under DSI. DSI covers all activities focused on advanced end-use efficiency and effective electricity utilisation, including demand response and energy efficiency.

5.5.1   Services provided by DSI

Demand-side resources[4] such as flexible loads, distributed generation and storage can provide various services to the power system by modifying the load consumption patterns. Such services can include load shifting, valley filling, peak clipping, dynamic energy management, energy efficiency improvement and strategic load growth [1, 12]. Simple daily domestic load profiles are used to illustrate the function of each service, as shown in Figures 5.13–5.16.

Load shifting is the movement of load between times of day (from on-peak to off-peak) or seasons. In Figure 5.13, a load such as a wet appliance (washing machine) that consumes 1 kW for 2 hours is shifted to off-peak time.

Figure 5.14 shows the main purpose of valley filling, which is to increase off-peak demand through storing energy, for example, in a battery of a plug-in electric vehicle or thermal storage in an electric storage heater. The main difference between valley filling and load shifting is that valley filling introduces new loads to off-peak time periods, but load shifting only shifts loads so the total energy consumption is unchanged (as shown in Figure 5.13).

Peak clipping reduces the peak load demand, especially when demand approaches the thermal limits of feeders/transformers, or the supply limits of the whole system. Peak clipping (Figure 5.15) is primarily done through direct load control of domestic appliances, for example, reducing thermostat setting of space heaters or control of electric water heaters or air-conditioning units. As peak clipping reduces the energy consumed by certain loads (in Figure 5.15, 2 kWh of energy is reduced), often consumers have to reduce their comfort.

Energy efficiency programs are intended to reduce the overall use of energy. Approaches include offering incentives to adopt energy-efficient appliances, lighting, and other end-uses; or strategies that encourage more efficient electricity use, for example, the feedback of

---

[4] End-use resources on the customer side of a meter that can be used to respond to the electric power system or market conditions.

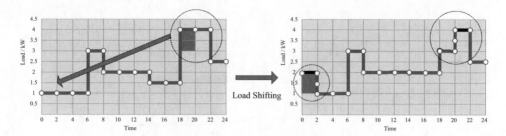

Figure 5.13 Load shifting

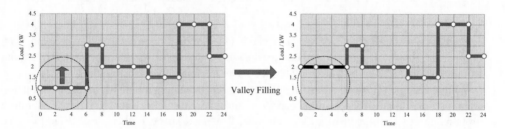

Figure 5.14 Valley filling

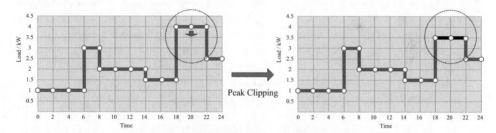

Figure 5.15 Peak clipping

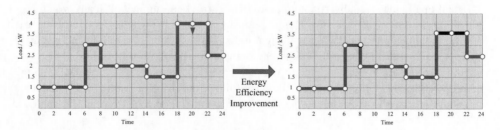

Figure 5.16 Energy efficiency improvement

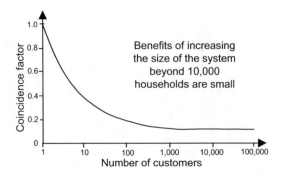

Figure 5.17   Reduction of coincidence factor with number of domestic customers

consumption and cost data to consumers, can lead to a reduction in total energy consumption. Figure 5.16 shows the reduction in energy demand when ten 60 W filament lamps (operating from 18.00 hrs to 22.00 hrs) are replaced by 20 W Compact fluorescent lamps.

With the deployment of smart metering and the development of home area automaton technologies, domestic appliances can be controlled in a more intelligent way, therefore bringing more flexibility to the demand side. The load shape is then flexible and can be controlled to meet the system needs. However, for the most effective DSI, the utility needs to know not only which loads are installed in the premises but which are in use. In this case two-way communication between the smart meter and network operators is necessary.

One way to de-carbonise the heating and transportation sectors is through electrification, for example, the increased utilisation of heat pumps and plug-in electric vehicles. However, the immediate consequence is the increase in electricity load causing potentially serious operational problems to both distribution and generation systems. Appropriate control and management are required if plant is not to be overloaded.

Demand-Side Integration describes a set of strategies which can be used in competitive electricity markets to increase the participation of customers in their energy supply. When customers are exposed to market prices, they may respond as described above, for example, by shifting load from the peak to the off-peak period, and/or by reducing their total or peak demand through load control, energy-efficiency measures or by installing distributed generation. Customers are able to sell energy services either in the form of reductions in energy consumption or through local generation.

Traditionally electric power systems were designed assuming that all loads would be met whenever the energy is requested. Domestic customers (and many other loads) use electricity at different times and this allows the design of the power system to benefit from diversity. For example, although a house in England may draw up to 10 kW at some times, its distribution supply system will be designed on an After Diversity Maximum Demand (ADMD) of only 1 or 2 kW. The coincidence of domestic demand follows the shape of Figure 5.17[5] and although distribution networks may only serve, say, 100 customers the transformers and cables also

---

[5] The coincidence factor is the ratio of the peak load drawn to the total connected load.

have significant thermal inertia and so a further reduction in the rating used for design may be assumed.

Demand-Side Integration has the potential to negate the beneficial effects of diversity. Consider a peak clipping control that sends a signal to switch off one hundred 3 kW water heaters that operate under thermostatic control. Although 100 water heaters have been installed, only, say, 20 will be drawing power at any one time. Thus the peak will be reduced by 60 kW. When, after, say, two hours, the water heaters are reconnected, all the water tanks will have cooled and a load of 300 kW will be reconnected. Thus DSI measures must consider both the disconnection of loads but also their reconnection and the payback of the energy that has not been supplied. It is much easier to manage both the disconnection of loads and their reconnection with bi-directional communications whereby the state of the loads can be seen by the control system.

Example 5.4

Consider the circuit shown in Figure 5.18. The 33/11 kV transformer has an on-load tap changer which maintains the load voltage at 11 kV. Calculate the percentage reduction in energy loss in the 33 kV line if load shifting shown in Figure 5.19 is managed. Ignore the 33/11 kV transformer losses.

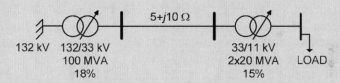

Figure 5.18   Figure for Example 5.4

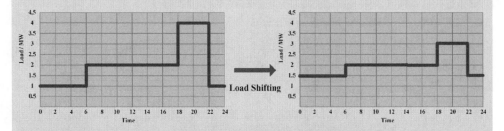

Figure 5.19   Load shifting scheme for Example 5.4

Answer

Table 5.5 shows the losses in the line when the load changes.

Table 5.5   Losses in the line when the load changes

Load (MW) P	Current in the 33 kV line (A) $I = P\big/\!\left(\sqrt{3} \times 33 \times 10^3\right)$	Losses (kW) $= I^2 R$
1	17.5	1.53
1.5	26.3	3.45
2	35.0	6.13
3	52.5	13.78
4	70.0	24.50

Energy loss without DSI $= 8 \times 1.53 + 12 \times 6.13 + 4 \times 24.5 = 183.8$ kWh
(Note that for 8 hrs per day the load was 1 MW, for 12 hrs per day it was 2 MW and for 4 hrs it was 4 MW.)
Energy loss with DSI $= 8 \times 3.45 + 12 \times 6.13 + 4 \times 13.78 = 156.3$ kWh
Percentage reduction $= [(183.8 - 156.3)/183.8] \times 100 = 15\%$.

5.5.2   Implementations of DSI

The implementations of DSI can be through price-based schemes or incentive-based schemes [13]. Price-based DSI encourages customer load changes in response to changes in the electricity price. Incentive-based DSI gives customers load modification incentives that are separate from, or in addition to, their retail electricity rates.

Various DSI programs are deployed and integrated within the power system core activities at different time scales of power system planning and operation, as shown in Figure 5.20.

5.5.2.1   Price-based DSI implementations

Tariffs and pricing can be effective mechanisms to influence customer behaviour, especially in unbundled electricity markets. Price schemes employed include time of use rates, real-time pricing and critical peak pricing:

- *Time of use (ToU)*: ToU rates use different unit prices for different time blocks, usually pre-defined for a 24-hour day. ToU rates reflect the cost of generating and delivering power during different time periods.
- *Real-time pricing (RTP)*: the electricity price provided by RTP rates typically fluctuates hourly, reflecting changes in the wholesale electricity price. Customers are normally notified of RTP prices on a day-ahead or hour-ahead basis.
- *Critical peak pricing (CPP)*: CPP rates are a hybrid design of the ToU and RTP. The basic rate structure is ToU. However, the normal peak price is replaced by a much higher CPP

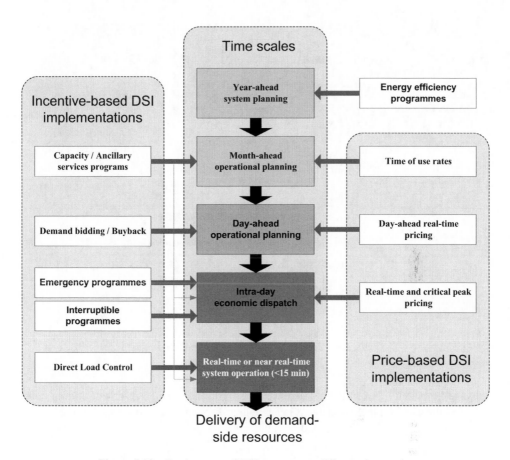

Figure 5.20 Deployment of DSI programs at different time scales

event price under predefined trigger conditions (for example, when system is suffering from some operational problem or the supply price is very high).

The various pricing schemes are illustrated in Figure 5.21. RTP defines hourly or half-hourly prices corresponding to changes in the intra-day or day-ahead cost of electricity generation and delivery. For RTP, one option is 'one-part' pricing, in which all use is priced at the hourly or spot price. Another approach is 'two-part' pricing. Two-part RTP tariff designs include a historical baseline of customer use, added to hourly prices only for marginal use above or below the baseline. Customers thus see market prices only at the margin. CPP uses real-time price at major system peaks. The CPP prices are restricted to a small number of hours per year, where electrical prices are much higher than normal peak prices, and their timing is unknown ahead of being called.

In price-based systems, the response of demand to price signals determines the DSI performance. Price elasticity is a measure used in economics to show the responsiveness, or elasticity, of the quantity demanded of a good or service to a change in its price. It gives the percentage

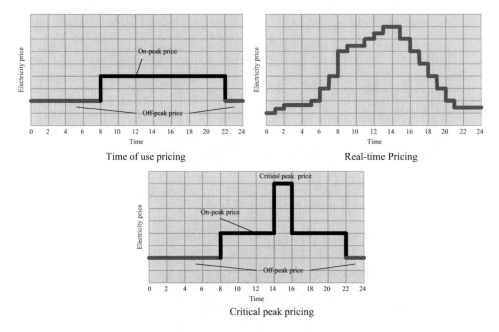

Figure 5.21 Illustration of various pricing schemes

change in quantity demanded in response to a 1 per cent change in price (holding constant all the other determinants of demand). There are different 'elasticities' in use to evaluate the available amount of the demand-side resources:

1. *Price elasticity of demand*: the percentage of change in demand as a result of a percentage of change in price (the elasticity should be a negative number)

$$\text{Price Elasticity} = \frac{\Delta D/D}{\Delta P/P}$$

where D is the demand, ΔD is the change in customer demand, P is the price, and ΔP is the change in the electricity energy price.

2. *Elasticity of substitution*: is a measure of the percentage change in the ratio of the peak to off-peak demand as a result of a percentage change in the ratio of the peak to off-peak price.

3. *Long-term price elasticity*: is the annual energy consumption response to an average change in energy price.

5.5.2.2 Incentive-based DSI implementations

Table 5.6 lists various kinds of implementations of incentive-based DSI.

Table 5.6   Implementations of incentive-based DSI

Implementations	Description
Direct load control	Customers' electrical appliances (e.g. air conditioner, water heater, space heating) are controlled remotely (for example, shut down or tuned by the controller) by the program operator on short notice
	Direct load control programmes are primarily offered to residential or small commercial customers
Interruptible/curtailable service	Curtailment options integrated into retail tariffs providing a rate discount or bill credit for agreeing to reduce load during system contingencies
	Penalties may be introduced for failing to curtail
	Interruptible programs have traditionally been offered only to the large industrial (or commercial) customers
Demand-side bidding/ Buy-back programs	Customers offer bids for curtailment based on wholesale electricity market prices
	Mainly offered to large customers (for example, one megawatt and over)
	For small customers, third parties (for example, aggregators) are needed to aggregate loads and bid in the market on behalf of them
Emergency demand response programs	Provide incentive payments to customers for load reduction during periods when the system is short of reserve
Capacity market programs	Customers offer load curtailment as system capacity to replace conventional generation
	Customers typically receive intra-day notice of curtailment events
	Incentives usually consist of upfront reservation payments, and penalties for customer failure to curtail
Ancillary services market programs	Customers bid load curtailments in ISO /RTO (Independent System Operator/Regional Transmission Organisation) markets as operational reserves
	If their bids are accepted, they are paid the market price for committing to be on standby
	If their load curtailments are needed, they are called by the ISO/RTO, and will be paid the spot market energy price

5.5.3   Hardware support to DSI implementations

The essential ICT infrastructure required for DSI can be provided by smart metering. In addition, load control switches, controllable thermostats, lighting controls and adjustable speed drives are required. Such equipment receives signals such as alarms or price signals and controls loads accordingly.

5.5.3.1   Load control switches

A load control switch is an electronic apparatus which consists of a communication module and a relay. It is wired into the control circuit of an air conditioning system, a water heater or a piece of thermal comfort equipment. The communication module is used to receive control signals from the DSI program operator (or a HAN). The time that the appliance will remain disconnected is generally pre-programmed (through an inbuilt clock).

5.5.3.2 Controllable thermostats

This type of apparatus combines a communication module with a controllable thermostat, and replaces conventional thermostats such as those on air conditioning systems or water heaters. The DSI program operator (or a HAN) can increase or decrease the temperature set point through the communication module, changing the functioning of the equipment and hence the electricity load.

Example 5.5

Consider an apartment having a thermal capacity factor $C = 0.4032$ kWh/K and a loss factor $L = 0.067$ kW/K. The heat required, Q, to maintain temperature at $\theta_H(t_1)$ is given by [14]:

$$Q = C \frac{[\theta_H(t_1) - \theta_H(t_0)]}{[t_1 - t_0]} + L \left[\frac{\theta_H(t_1) + \theta_H(t_0)}{2} - \theta_A \right]$$

where the temperature inside the flat changes from $\theta_H(t_0)$ to $\theta_H(t_1)$ within time period t_0 to t_1 and the outside temperature remains constant at θ_A. Take $t_1 - t_0 = 1$ hr.

If $\theta_A = 0\,°C$ and the flat has an electric heater, calculate the reduction in energy usage within an hour if the temperature within the flat was reduced from 20 °C to 19 °C using an automatic thermostat when compared to an hour where temperature is maintained at 20 °C.

Answer

If temperature is maintained at 20 °C

$$\theta_H(t_0) = \theta_H(t_1) = 20\,°C$$

$$Q = 0.067 \left[20° - 0° \right] = 1.34\,kW$$

If the temperature was reduced from 20 °C to 19 °C.

$$\theta_H(t_0) = 20\,°C \text{ and } \theta_H(t_1) = 19\,°C$$

$$Q = 0.4032 \times \frac{[19° - 20°]}{1} + 0.067 \times \left[\frac{19° + 20°}{2} - 0° \right]$$

$$= 0.9\,kW$$

Therefore the reduction of energy consumption for one hour $= (1.34{-}0.9) = 0.44$ kW or 0.44 kWh. Note that only part of this reduction would be maintained for a second hour at 19 °C.

Table 5.7   Lighting control strategies

Strategy	Description	Estimated energy savings
Planning program	Elimination or reduction of lighting during periods of low occupancy	10–30% with programmers 30–60% with personnel detectors
Natural lighting	Deactivation or dimming of lighting according to the natural lighting in the building	10–15% for deactivation 15–35% for dimming
Constant light levels	Efficient compensation of low levels of natural light	10–15%
Tuning	Tuning the level of lighting according to the needs of the area	10–20%
Load shedding	Temporary lighting reduction to reduce peak demand	—
Light compensation	Modification of the level of lighting for increased visual comfort	—

5.5.3.3   Lighting control

Lighting control equipment is used to manage the energy used by lighting in a more efficient way. Lighting control strategies for energy consumption reduction are listed in Table 5.7. Estimated energy savings are presented for each case. These savings are based upon estimated average consumption, the time of use and user behaviour.

5.5.3.4   Adjustable speed drives

Adjustable Speed Drives (ASDs) allow electric motors driving pumps, ventilation units and compressors to function over a continuous speed range. The loads of the majority of motorised appliances change over time and equipment is often operated at less than full load. ASDs allow the motors to satisfy the required functioning conditions and to economise power and energy use when the system is not functioning at its maximum load. Directly connected motors for pump and fans are often oversized and the fluid flow throttled for control. Replacement of this system by an ASD can yield considerable saving of energy.

5.5.4   Flexibility delivered by prosumers from the demand side

Some customers have installed Distributed Generation (DG) and energy storage (these are generally referred to Distributed Energy Resources – DERs) in their premises. Such customers not only consume electricity, but also are able to manage their capacity to supply power to the grid. Hence they are also called prosumers. The controllability of active power is fundamental for the commercial integration of prosumers. The controllability of reactive power of distributed energy resources has a great influence on the fault ride-through capability and the provision of ancillary services.

The flexibility provided by prosumers depends on the level of DER penetration, the DER technologies utilised, and the locations and interfacing technologies of the connection points with the power network. DER connected to low and medium voltage levels will have more opportunities to provide local network services than generation connected at higher voltage distribution networks.

5.5.5 System support from DSI

Emergency load shedding has been used in many power systems to maintain the integrity of the power system in the event of a major disturbance. It is triggered by under-frequency relays when the frequency drops under a certain threshold, for example, 48.8 Hz in England and Wales, and consists of the tripping of entire distribution feeders. Load shedding is planned by the TSO but is implemented by the DNOs who arrange the tripping of distribution feeders and choose which feeders are tripped.

During normal operation, the GB Transmission System Operator (NGET) maintains the frequency at 50 ± 0.2 Hz. In order to maintain frequency, NGET buys frequency response services. When the frequency goes up, high frequency response is used to reduce the power output of the large generators and hence the frequency. A sudden drop in frequency is contained using primary response (Figure 5.22). This should be delivered within 10 seconds and maintained for another 20 seconds [15]. The system frequency is brought back to normal using secondary response which lasts from 30 seconds to 30 minutes. If the frequency continues to drop below 48.8 Hz, demand is disconnected (load shedding) to prevent shutdown of the power system [16].

Primary and secondary response is usually provided by partially loaded generators increasing their output. However, this incurs significant costs as de-loading a steam turbine set by, for example, 100 MW only provides around 20 MW of primary response. Primary response requires energy to be used that is stored as high pressure steam in the boiler drum and so only a fraction of the de-loaded power can be used. Operating a steam turbine set at below its maximum output significantly reduces its thermal efficiency.

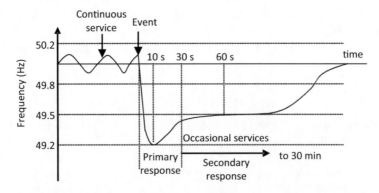

Figure 5.22   Frequency control in England and Wales [15]

DSI programs can significantly reduce the requirement for primary and secondary responses from partially loaded generators by shedding load in a controlled manner. Large loads that are contracted to provide frequency response are typically steelworks or aluminium smelters though hospitals and banks that have their own generators can also take part in this market. Using load in this way reduces the system operating cost and, depending on the alternative generation being used, CO_2 emissions.

Recent consultations issued in the UK indicate that smart meters will have the capability for Direct Load Control (DLC)[6] [17]. DLC directly switches off loads to balance supply and demand during emergency conditions such as sudden loss of generation. In the USA, DLC has been used by power system operators or by companies which provide DSI services [18].

References

[1] Kreith, F. and Goswami, D.Y. (2007) *Handbook of Energy Efficiency and Renewable Energy*, CRC Press, New York.

[2] Ziegler, S., Woodward, R.C., Iu, H.H.-C. and Borle, L.J. (2009) Current sensing techniques: A review. *IEEE Sensors Journal*, **9**(4), 354–376.

[3] Millman, J. and Grabel, A. (1987) *Microelectronics*, McGraw-Hill Book Company, Singapore.

[4] Allen, P.E. and Holberg, D.R. (2002) *CMOS Analog Circuit Design*, Oxford University Press, New York.

[5] Kester, W. (2009) *ADC Architectures III: Sigma-Delta ADC Basics*, Analogue Devices, MT-022 Tutorial, 2009, http://www.analog.com/static/imported-files/tutorials/MT-022.pdf (accessed on 4 August 2011).

[6] Webster, J.G. (1998) *The Measurement, Instrumentation and Sensors Handbook*, CRC Press, New York.

[7] Kushiro, N., Suzuki, S., Nakata, M. *et al.* (2003) Integrated residential gateway controller for home energy management system. *IEEE Transactions on Consumer Electronics*, **49**(3), 629–636.

[8] Young, S. and Stanic, R. (2009) *SmartMeter to HAN Communications*, SmartGrid Australia Intelligent Networking Working Group, July 2009, http://smartgridaustralia. com.au/SGA/Documents/IN_Work_Group_SmartMeter_HAN_Comms.pdf (accessed on 8 August 2011).

[9] *Smart Metering Implementation Programme: Statement of Design Requirements*, Ofgem E-Serve, July 2010, http://www.decc.gov.uk/assets/decc/consultations/smart-meter-imp-prospectus/225-smart-metering-imp-programme-design.pdf (accessed on 4 August 2011).

[10] *Report on Regulatory Requirements*, EU project OPENmeter, July 2009, http:// www.openmeter.com/files/deliverables/Open_Meter_D1.2_Regulation_v1.1_20090717. pdf (accessed on 4 August 2001).

---

[6] A DSI activity which remotely shuts down or changes thermostat setting of customer's electrical equipment (for example, air conditioning, water heater) on short notice.

[11] De Craemer, K. and Deconinck, G. (2010) *Analysis of State-of-the-Art Smart Metering Communication Standards*, https://lirias.kuleuven.be/bitstream/123456789/265822/1/SmartMeteringCommStandards.pdf (accessed on 4 August 2011).

[12] *IEA Demand Side Management Programme*, http://www.ieadsm.org/ (accessed on 4 August 2011).

[13] Gellings, C.W. (2009) *The Smart Grid: Enabling Energy Efficiency and Demand Response*, The Fairmont Press, Lilburn.

[14] Chaudry, M., Ekanayake, J. and Jenkins, N. (2008) Optimum control strategy of a mCHP unit. *International Journal of Distributed Energy Resources*, **4**(4), 265–280.

[15] Erinmez, I.A., Bickers, D.O., Wood, G.F. and Hung, W.W. (1999) NGC experience with frequency control in England and Wales: provision of frequency response by generators. *IEEE Power Engineering Society Winter Meeting*, **1**, 590–596.

[16] *Report of the Investigation into the Automatic Demand Disconnection Following Multiple Generation Losses and the Demand Control Response that Occurred on the 27th May 2008*, http://www.nationalgrid.com/NR/rdonlyres/D680C70A-F73D-4484-BA54-95656534B52D/26917/PublicReportIssue1.pdf (accessed on 4 August 2011).

[17] Department of Energy and Climate Change, UK, *Smart Metering Implementation Programme: Prospectus*, March 2011, http://www.decc.gov.uk/en/content/cms/consultations/smart_mtr_imp/smart_mtr_imp.aspx (accessed on 4 August 2011).

[18] Federal Energy Regulatory Commission, USA, *Assessment of Demand Response and Advanced Metering 2007 Staff Report*, Sept. 2007, http://www.ferc.gov/legal/staff-reports/09-07-demand-response.pdf (accessed on 4 August 2011).

6

Distribution Automation Equipment

6.1 Introduction

Modern electric power systems are supplied by large central generators that feed power into a high voltage interconnected transmission network. The power, often transmitted over long distances, is then passed down through a series of distribution transformers to final circuits for delivery to customers (Figure 6.1, also refer to Plate 1).

Operation of the generation and transmission systems is monitored and controlled by Supervisory Control and Data Acquisition (SCADA) systems. These link the various elements through communication networks (for example, microwave and fibre optic circuits) and connect the transmission substations and generators to a manned control centre that maintains system security and facilitates integrated operation. In larger power systems, regional control centres serve an area, with communication links to adjacent area control centres. In addition to this central control, all the generators use automatic local governor and excitation control. Local controllers are also used in some transmission circuits for voltage control and power flow control, for example, using phase shifters (sometimes known as quadrature boosters).

Traditionally, the distribution network has been passive with limited communication between elements. Some local automation functions are used such as on-load tap changers and shunt capacitors for voltage control and circuit breakers or auto-reclosers for fault management. These controllers operate with only local measurements and wide-area coordinated control is not used.

Over the past decade, automation of the distribution system has increased in order to improve the quality of supply and allow the connection of more distributed generation. The connection and management of distributed generation are accelerating the shift from passive to active management of the distribution network. Network voltage changes and fault levels are increasing due to the connection of distributed generation [1]. Without active management of the network, the costs of connection of distributed generation will rise and the connection of additional distributed generation may be limited [2].

Smart Grid: Technology and Applications, First Edition.
Janaka Ekanayake, Kithsiri Liyanage, Jianzhong Wu, Akihiko Yokoyama and Nick Jenkins.
© 2012 John Wiley & Sons, Ltd. Published 2012 by John Wiley & Sons, Ltd.

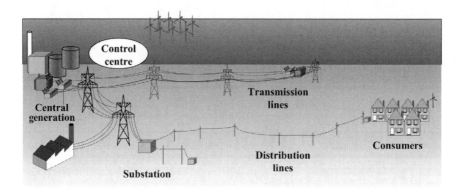

Figure 6.1   Typical power system elements

The connection of large intermittent energy sources and plug-in electric vehicles will lead to an increase in the use of Demand-Side Integration and distribution system automation.

6.2   Substation automation equipment

The components of a typical legacy substation automation system are shown in Figure 6.2. Traditionally, the secondary circuits of the circuit breakers, isolators, current and voltage transformers and power transformers were hard-wired to relays. Relays were connected with multi-drop serial links to the station computer for monitoring and to allow remote interrogation.

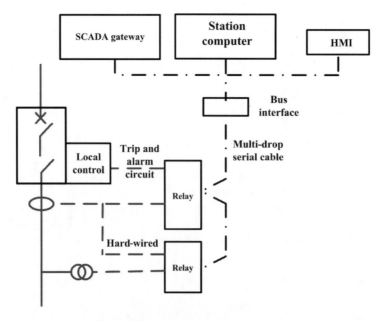

Figure 6.2   Substation components [3, 4]

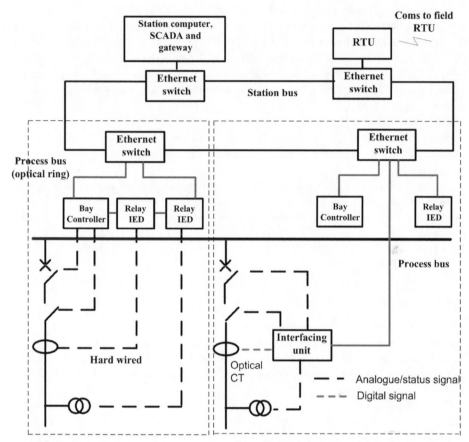

Connection 1: Secondary circuits of field equipment vare hard wired to relay IEDs and bay controller. The process bus is ring connected

Connection 2: Secondary circuits of field equipment are hard wired to interfacing unit. The process bus is star connected

Figure 6.3   A modern substation [3,5]

However, the real-time operation of the protection and voltage control systems was through hard-wired connections.

The configuration of a modern substation automation system is illustrated in Figure 6.3. Two possible connections (marked by boxes) of the substation equipment are shown in Figure 6.3. Although it may vary from design to design, generally it comprises three levels:

- The *station level* includes the substation computer, the substation human machine interface (which displays the station layout and the status of station equipment) and the gateway to the control centre.
- The *bay level* includes all the controllers and intelligent electronic devices (which provide protection of various network components and a real-time assessment of the distribution network).
- The *process level* consists of switchgear control and monitoring, current transformers (CTs), voltage transformers (VTs) and other sensors.

In connection 1, analogue signals are received from CTs and VTs (1 A or 5 A and 110 V) as well as status information and are digitised at the bay controller and IEDs. In connection 2, analogue and digital signals[1] received from CTs and VTs are digitised by the interfacing unit. The process bus and station bus take these digital signals to multiple receiving units, such as IEDs, displays, and the station computer that are connected to the Ethernet network. To increase reliability, normally two parallel process buses are used (only one process bus is shown in Figure 6.3) [5].

The station bus operates in a peer-to-peer mode. This bus is a LAN formed by connecting various Ethernet switches through a fibre-optic circuit. The data collected from the IEDs is processed for control and maintenance by SCADA software that resides in the station computer.

The hard-wiring of traditional substations required several kilometres of secondary wiring in ducts and on cable trays. This not only increased the cost but also made the design inflexible. In modern substations as inter-device communications are through Ethernet and use the same communication protocol, IEC 61850, both the cost and physical footprint of the substation have been reduced.

6.2.1   Current transformers

The normal load current of transmission and distribution circuits varies up to hundreds or even thousands of amperes. When a short circuit fault occurs, the current may increase to more than 20 times the normal load current. Current transformers (CTs) are used to transform the primary current to a lower value (typically 1 or 5 A maximum) suitable for use by the IEDs or interfacing units.

The majority of CTs, which are now in service, are iron cored with a secondary winding on the core. The primary is often the main circuit conductor forming a single turn. The operating principle of these transformers can be found in [3, 6, 7]. The iron core of these transformers introduces inaccuracies in the measurements due to the presence of magnetising current (which only appears on the primary), flux leakage, magnetic saturation and eddy current heating. In order to minimise their measurement errors, the design is optimised for the specific application.

Measurement CTs are used to drive ammeters, power and energy meters. They provide accurate measurements up to 120 per cent of their rated current. In contrast, protection CTs provide measurement of the much greater fault current and their accuracy for load current is generally less important.

Measurement CTs are specified by IEC 60044-1 according to their accuracy classes, of 0.1, 0.2, 0.5 and 1 per cent at up to 120 per cent of rated current.

Protection CTs are normally described for example, as '10 VA Class 10P 20'. The first term (10 VA) is the rated burden of the CT that can have a value of 2.5, 5, 10, 15 or 30 VA. The accuracy class (10P) defines the specified percentage accuracy. The last term (20) is the accuracy limit (the multiple of primary current up to which the CT is required to maintain its specified accuracy with rated burden connected). The accuracy limit can be 5, 10, 20 or 30.

---

[1] Optical CTs and VTs provide digital signals whereas conventional CTs and VTs provide an analogue signal (see Sections 6.2.1 and 6.2.2).

Class 10P is designated in ANSI/IEEE C57.13 as class C where the CT is classified by 'C' followed by a number. This number indicates the secondary terminal voltage that the transformer can deliver to a standard burden at 20 times the rated current without exceeding an accuracy of 10 per cent.

There are other classes of CTs such as Class T and X of IEEE C57.13, and Classes 3, 5 and PX of IEC 60044-1. More details about these current transformers can be found in the respective standards [8, 9].

Example 6.1

Part of a distribution circuit is shown in Figure 6.4. The CT used is a 10 VA Class 10P 20 and has a secondary resistance of 0.6 Ω and a magnetising reactance of $j15$ Ω. Using the equivalent circuit of the CT [3, 10], obtain the percentage current magnitude error and phase displacement error for the rated current and for the current at the accuracy limit when the rated burden is connected to the secondary.

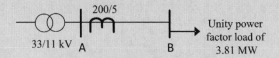

Figure 6.4   Figure for Example 6.1

Answer

As the rated burden is 10 VA, the corresponding secondary resistance is given by

$$R_{\text{burden}} = \frac{10}{5^2} = 0.4\,\Omega$$

The per-phase equivalent circuit of the CT with rated current flowing is shown in Figure 6.5:

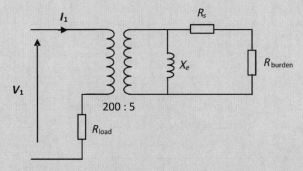

Figure 6.5   Per-phase equivalent circuit of the CT

$$V_1 = \frac{11000}{\sqrt{3}} = 6350.9 \text{ V}$$

$$R_{load} = \frac{3 \times (6350.9)^2}{3.81 \times 10^6} = 31.75 \ \Omega$$

Transforming the primary quantities into secondary (multiplying voltage by the turns ratio, that is, 40 and primary resistance by 40^2) the following circuit is obtained, as shown in Figure 6.6:

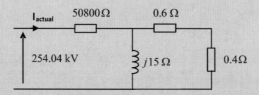

Figure 6.6   Secondary referred equivalent circuit of Figure 6.5

$$\text{Actual current in the secondary circuit} = \frac{254.04 \times 10^3}{50800 + [j15 \times 1.0/(1.0 + j15)]}$$

$$= 5.0007\angle 0°$$

$$\text{Therefore the percentage current magnitude error} = \frac{5.0007 - 5}{5} \times 100 = 0.014\%$$

Phase angle error is $0°$.

When the current is at the accuracy limit, that is, 200×20 A is flowing in the primary circuit, instead of the R_{load} in the equivalent circuit, R_{fault} appears in the transformer primary.

$$R_{fault} = \frac{6350.9}{4000} = 1.59 \ \Omega$$

$$\text{New actual current in the secondary circuit} = \frac{254.04 \times 10^3}{1.59 \times 40^2 + [j15 \times 1.0/(1.0 + j15)]}$$

$$= 99.8194\angle - 0.0015°$$

Therefore, the percentage current magnitude error =

$$\frac{99.8194 - 5 \times 20}{5 \times 20} \times 100 = -0.18\%$$

The phase angle error is $0.0015°$.

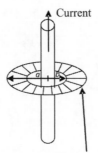

A wire wound on a non-magnetic core

Figure 6.7 A Rogowski coil

As shown in connection 2 of Figure 6.3 (normally used in modern substations), the current measurements from a CT are digitised and made available to the process bus and used by a number of devices. Multiple use of the same digitised measurement requires high accuracy CTs that measure both load and fault currents.

While iron cored CTs and hybrid CTs (an iron cored CT with an optical transmitter) remain the most widely used CTs in the power system, high accuracy designs such as the Rogowski coil formed on a printed circuit board and optical CTs [3, 11, 12] are becoming available.

Rogowski coil CTs are used in a Gas Insulated Substations (GIS). The secondary winding of the coil is a multi-layer printed circuit board as shown in Figure 6.7 with the upper and lower tracks of each layer connected by metal vias (thus forming a rectangular coil). The voltage induced on the secondary windings due to primary current (see Box 6.1) is integrated (as a Rogowski coil gives di/dt) by the sensor electronics to obtain the value of the primary current.

Box 6.1 A Rogowski coil

The Rogowski coil detects the magnetic field created by the change in conductor current and generates a voltage (v) proportional to it.

$$v = \frac{\mu NL}{2\pi} \ln \left| \frac{a}{b} \right| \frac{di}{dt}$$

where N is number of turns, L is thickness of the printed circuit board (multi-layers are used), and di/dt is the change in primary current.

Optical CTs use the Faraday effect (Box 6.2), whereby the plane of polarisation of a light beam when subjected to a magnetic field, is rotated through an angle. This angle of rotation is proportional to the magnetic field thus to the primary current. Figure 6.8 shows this type

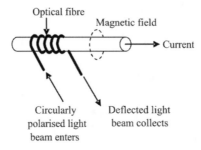

Figure 6.8   A simple optical CT

of CT in its simplest form. The opto-electronics compares the polarisation of the light beam entered into the optical fibre and that collected after being subjected to the circular magnetic field. The angle of deflection is used to generate digital signals proportional to the line current.

Box 6.2   The Faraday Effect

The Faraday effect describes an interaction between light and a magnetic field in a medium. A polarised light beam rotates when subjected to a magnetic field (Figure 6.9). The rotation of the plane of polarisation is proportional to the intensity of the magnetic field in the direction of the beam of light.

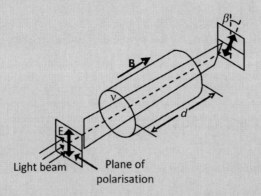

Figure 6.9   The Faraday effect

   The angle of rotation β in radians is given by $\beta = v\mathbf{B}d$ where $\mathbf{B}$ is the magnetic flux density (in T), d is the length of the path (in m) and v the Verdet constant for the material.

   Some other designs are being developed based on the Faraday effect that use a disc of an optically active material around the conductor [3, 7, 11]. The light enters the disc from one side and travels around it (thus around the conductor) and is collected at the other end.

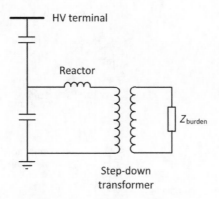

Figure 6.10 A high voltage CVT

6.2.2 Voltage transformers

It is necessary to transform the power system primary voltage down to a lower voltage to be transferred through process bus to IEDs, bay controller and station computer. The secondary voltage used is usually 110 V. At primary voltages up to 66 kV, electromagnetic voltage transformers (similar to a power transformer with much lower output rating) are used but at 132 kV and above, it is common to use a capacitor voltage transformers (CVT).

As the accuracy of voltage measurements may be important during a fault, protection and measuring equipment are often fed from the same voltage transformer (VT). IEC 60044-2 and ANSI/IEEE C57.13 define the accuracy classes of VTs. Accuracy classes such as 0.1, 0.2, 0.5, 1.0 and 3.0 are commonly available. For example, Class 0.1 means the percentage voltage ratio error should not exceed 0.1 per cent at any voltage between 80 and 120 per cent of rated voltage and with a burden of between 25 and 100 per cent of rated burden.

The basic arrangement of a high voltage CVT is a capacitor divider, a series reactor (to compensate for the phase shift introduced by the capacitor divider) and a step-down transformer (for reducing the voltage to 110 V). The voltage is first stepped down to a high value by a capacitor divider and further reduced by the transformer, as shown in Figure 6.10 [3, 6, 7].

For applications up to 11 kV, optical CVTs are now available. Due to the lower voltage involved the inductor and transformer (in Figure 6.10) are replaced by an opto-electronic circuit mounted on the base tank (see Figure 6.11). In this arrangement there is no L-C circuit to resonate, and hence no oscillations, over-voltages or any possibility of ferro-resonance.

Some VTs use a similar technique to optical CTs based on the Faraday effect. In this case, an optical fibre is situated inside the insulator running from top to bottom and is fed by a circular polarised light signal. Due to the magnetic field between the HV terminal and the base tank, the polarisation of the light signal changes and that deflection is used to obtain the HV terminal voltage.

6.2.3 Intelligent electronic devices

The name Intelligent Electronic Device (IED) describes a range of devices that perform one or more of functions of protection, measurement, fault recording and control. An IED consists

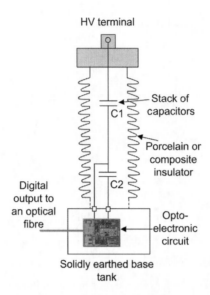

HV terminal

C1 — Stack of capacitors

Porcelain or composite insulator

C2

Digital output to an optical fibre

Opto-electronic circuit

Solidly earthed base tank

Figure 6.11   Basic circuit of an optical CVT.

of a signal processing unit (as discussed in Sections 5.3.2 and 5.3.3), a microprocessor with input and output devices, and a communication interface. Communication interfaces such as EIA 232/EIA 483, Ethernet, Modbus and DNP3 are available in many IEDs.

6.2.3.1   Relay IED

Modern relay IEDs combine a number of different protection functions with measurement, recording and monitoring. For example, the relay IED shown in Figure 6.12 has the following protection functions:

- three-phase instantaneous over-current: Type 50 (IEEE/ANSI designation);
- three-phase time-delayed over-current (IDMT): Type 51;
- three-phase voltage controlled or voltage restrained instantaneous or time-delayed over-current: Types 50V and 51V;
- earth fault instantaneous or time-delayed over-current: Types 50N and 51N.

The local measurements are first processed and made available to all the processors within the protection IED. A user may be able to read these digitised measurements through a small LED display as shown in Figure 6.13. Furthermore, a keypad is available to input settings or override commands.

Various algorithms for different protection functions are stored in a ROM. For example, the algorithm corresponding to Type 50 continuously checks the local current measurements against a set value (which can be set by the user or can be set remotely) to determine whether there is an over-current on the feeder to which the circuit breaker is connected. If the current is greater than the setting, a trip command is generated and communicated to the Circuit

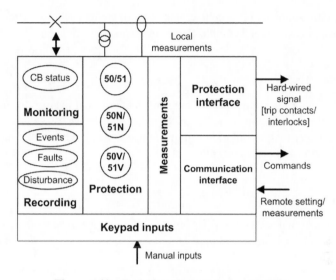

Figure 6.12 Typical configuration of a relay IED

Figure 6.13 Relay IED. *Source:* Courtesy of Toshiba

Breaker (CB). IEDs have a relay contact that is hard-wired (in series) with the CB tripping coil and the tripping command completes the circuit, thus opening the CB.

6.2.3.2 Meter IED

A meter IED provides a comprehensive range of functions and features for measuring three-phase and single-phase parameters. A typical meter IED measures voltage, current, power, power factor, energy over a period, maximum demand, maximum and minimum values, total harmonic distortion and harmonic components.

6.2.3.3 Recording IED

Even though meter and protection IEDs provide different parameters (some also have a data storage capability), separate recording IEDs are used to monitor and record status changes in the substation and outgoing feeders.

Continuous event recording up to a resolution of 1 ms is available in some IEDs. These records are sometimes interrogated by an expert to analyse a past event. This fault recorder records the pre-fault and fault values for currents and voltages. The disturbance records are used to understand the system behaviour and performance of related primary and secondary equipment during and after a disturbance.

6.2.4 Bay controller

Bay controllers (Figure 6.14) are employed for control and monitoring of switchgear, transformers and other bay equipment. The bay controller facilitates the remote control actions (from the control centre or from an on-site substation control point) and local control actions (at a point closer to the plant).

The functionalities available in a bay controller can vary, but typically include:

- CB control
- switchgear interlock check
- transformer tap change control
- programmable automatic sequence control.

6.2.5 Remote terminal units

The distribution SCADA system acquires data (measurements and states) of the distribution network from Remote Terminal Units (RTU). This data is received by an RTU situated in the substation (referred to here as the station RTU), from the remote terminal units situated in other parts of the distribution network (referred to here as the field RTU).

The field RTUs act as the interface between the sensors in the field and the station RTU. The main functions of the field RTU are to: monitor both the analogue and digital sensor signals (measurements) and actuator signals (status), and convert the analogue signals coming from the sensors and actuators into digital form. The station RTU acquires the data from the field

Figure 6.14 Bay controller. *Source:* Courtesy of Toshiba

RTUs at a predefined interval by polling. However, any status changes are reported by the field RTUs whenever they occur.

Modern RTUs, which are microprocessor-based, are capable of performing control functions in addition to data processing and communication. The software stored in the microprocessor sets the monitoring parameters and sample time; executes control laws; sends the control actions to final circuits; sets off calling alarms and assists communications functions. Some modern RTUs have the capability to time-stamp events down to a millisecond resolution.

6.3 Faults in the distribution system

When a fault occurs in the transmission or distribution system, the power system voltage is depressed over a wide area of the network and only recovers when the fault is cleared. Transmission systems use fast-acting protection and circuit breakers to clear faults within around 100 ms. In contrast, the time-graded over-current protection of distribution circuits and their slower CBs only clear faults more slowly, typically taking up to 500 ms.

Fast clearance of faults is important for industrial, commercial and increasingly for domestic premises. Many industrial processes rely on motor drives and other power electronic equipment which is controlled by microprocessors. Commercial and domestic premises use ever more Information Technology Equipment (ITE). This equipment is becoming increasingly sensitive to voltage dips [13, 14].

Figure 6.15 shows the well-known ITI (CBEMA) curve which specifies the AC voltage envelope that can be tolerated by Information Technology Equipment (note the log scales on

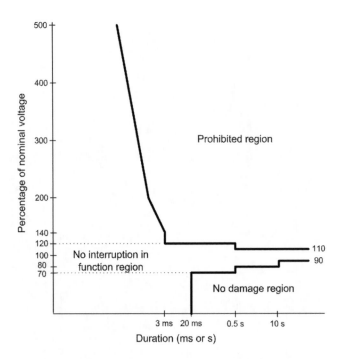

Figure 6.15   ITI (CBEMA) Curve (Revised 2000) [15]

the axes). During a fault on the AC network, depending on the location of the fault, the voltage will drop. The subsequent operation of the ITE depends on the fault clearance time and the voltage dip. For example, for a fault that creates a 40 per cent voltage dip (60 per cent retained voltage) for 400 ms, there is no damage to the ITE but its normal operation is not expected. However, for a fault that creates a 20 per cent voltage dip (80 per cent retained voltage) for 400 ms, the ITE should work normally.

Short-circuit faults are inevitable in any distribution system and so interruption in function of sensitive load equipment can only be avoided by doing the following:

- ensuring the load equipment is robust against these transient voltage changes;
- using very high speed protection and circuit breakers;
- adding equipment to mitigate the voltage depressions for example, a Dynamic Voltage Restorer (DVR) or STATCOM [14, 16].

A sustained electricity outage may lead to severe disruption and economic loss, especially for industrial processes. Hence many Regulators impose penalties for the loss of electricity to customers (in the UK for interruptions over 3 minutes).

In these circumstances, distribution network operators are concerned to increase the speed of isolating the fault and restoring supply. Increasingly they are applying automatic supply restoration techniques which use automatic reclosers, remotely controlled switches, remote measurements and sometimes local Agents (a piece of software running in a local computer).

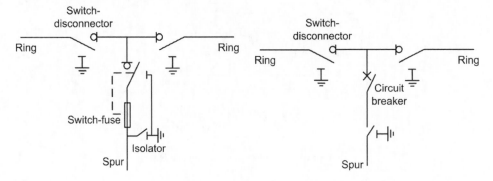

Figure 6.16   Typical RMU configurations

6.3.1   Components for fault isolation and restoration

Whenever there is a fault on a part of the distribution network, the fault current should be interrupted rapidly, the faulted section isolated from the healthy network and, then once the fault has been removed, supplies to customers should be restored. This is achieved through a range of equipment generally known as switchgear. The term switchgear includes: a circuit breaker which is capable of making and breaking fault currents, a recloser which is essentially a CB with a limited fault-breaking capacity and variable pattern of automatic tripping and closing, a switch disconnector which has a limited fault-making capability (and which is capable of making and breaking normal load current), and a sectionaliser which is capable of making and breaking normal load current but not the fault current.

33 kV substation switchgear may be located inside the substation building or in an outdoor high voltage compound; whereas 11 kV substation switchgear is normally indoor type or mounted on the overhead line poles. The protection and metering equipment of substation switchgear is housed inside the substation control room. Out on 11 kV circuits, away from primary substations, switchgear is usually outdoors, often as a Ring Main Unit (RMU) [17, 18]. A typical RMU consists of two switch-disconnectors and a switch-fuse[2] or circuit breaker (Figure 6.16).

In the past, switch-disconnectors in an RMU required manual operation but automatic operation can now be achieved for these designs by retrofitting an actuator mechanism. The most commonly used retrofit actuator mechanism is a motor wound spring. Remote switching is initiated from a field RTU[3] as shown in Figure 6.17. For clarity, only switch-disconnectors are shown in Figure 6.17. The fuse switch either acts directly on the switch-disconnect trip bar or is automated by connecting a shunt trip coil.

In modern switchgear, instead of a motor wound spring, compressed gas or magnetic actuators are increasingly used [17].

---

[2] At the moment the fuse operates, a striker pin located in the fuse cap is ejected to trip the spring-loaded switch-disconnector.

[3] In the past, a field RTU was only capable of monitoring the sensor (measurements) and actuator (status) signals and transmitting them to the station RTU. Therefore a programmable logic controller (which was hard-wired to the RTU) was used to control the motorised switchgear. Modern RTUs, which are microprocessor-based, are capable of performing control functions in addition to data processing and communication.

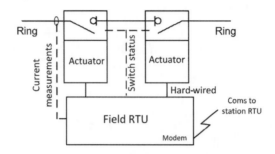

Figure 6.17   Part of an automated RMU

Most overhead line faults are transient and self-clearing once the circuit is de-energised. Hence either auto-reclose of the substations CBs or a self-controlled recloser, which can perform a variable pattern of tripping and reclosing, is used on many overhead distribution circuits. These will prevent unnecessary sustained outages for temporary faults. Most reclosers have instantaneous and time-delayed tripping characteristics. Standard practice is to use a number of instantaneous trips followed by time-delayed trips. The upper sequence of Figure 6.18 shows how a recloser clears a temporary fault after two recloses. The purpose of the delayed tripping (see the lower sequence of Figure 6.18) is if the fault is permanent, the delay allows sufficient current (and time) to pass through the downstream fuse so as to clear the fault. In

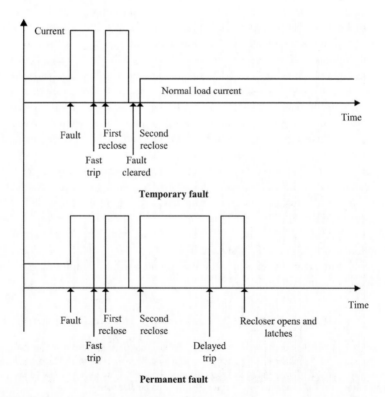

Figure 6.18   Reclose sequence for temporary and permanent faults. *Note:* Fault current is not to scale

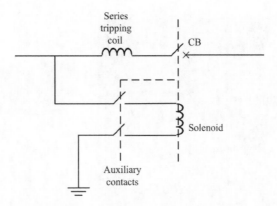

Figure 6.19 Recloser arrangement. *Note:* Only one phase is shown [7, 18]

some instances the delay time allows the fault to burn itself out so that a subsequent reclose restores the supply.

Pole-mounted reclosers are widely used in distribution circuits. They have different voltage ratings (for example, 11, 15, 33 kV) and interrupting currents of 8 to 16 kA. The energy required to operate the reclosing arrangement is provided by a solenoid as shown in Figure 6.19. In this arrangement, whenever the line current is high (due to a fault), the series trip coil opens the vacuum CB. As the CB is opened, the auxiliary contacts are closed automatically, thus providing energy for the reclosing operation.

A distribution feeder which employs a pole-mounted recloser is shown in Figure 6.20. The recloser characteristic is selected so as to make sure that its fast operating time is much faster than the operating time of the downstream fuses and its slow operating time is slower than the operating time of the fuses (see Figure 6.23).

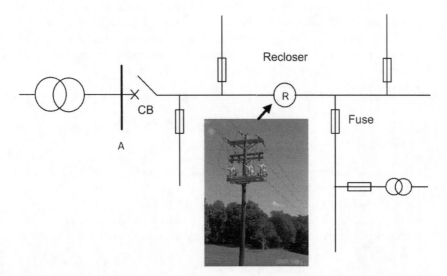

Figure 6.20 A typical distribution feeder with recloser and fuses. *Source:* Courtesy of S & C Electric Europe Ltd.

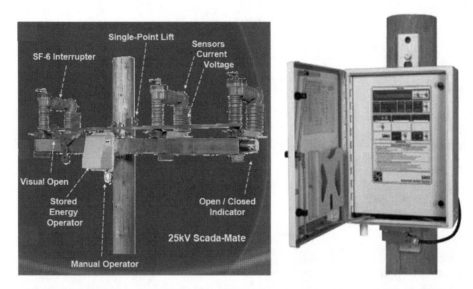

Figure 6.21   A pole-mounted recloser and RTU. *Source:* Courtesy of S & C Electric Europe Ltd.

An RTU can be incorporated with a pole-mounted recloser for remote switching and changing its settings remotely (see Figure 6.21, also refer to Plate 2). In the USA, alternative trip settings are used during storms to save the fuses as it is difficult to reach remote areas for fuse replacement [19].

A sectionaliser is an automatic isolator which can only be used to isolate a section of a distribution circuit once a fault is cleared by an upstream recloser. When the recloser opens, a self-powered control circuit in the sectionaliser increments its counter. The recloser is reclosed after a short delay to see whether the fault is temporary. If the fault is temporary and cleared by the recloser, the sectionaliser resets its counter and comes back to its normal state. However, if the fault is permanent, when the number of counts in the sectionaliser reaches a pre-defined number, it opens and isolates the downstream section of circuit. If the fault is downstream of the sectionaliser, the recloser then restores supply to the upstream section. This minimises the number of consumers affected by a permanent fault, and a more precise indication of the fault location is provided.

Example 6.2

Two distribution networks are shown in Figure 6.22. Network A has fuses to protect each line section whereas network B has sectionalisers. The characteristic of fuse F4 (solid curve) and that of the pole-mounted recloser (dotted curve) is shown in Figure 6.23. The sectionaliser has a pick-up current of 200 A. Discuss the operation of the protection arrangements in the two circuits for a temporary fault (as shown) that produces a fault current of 1000 A and 2000 A.

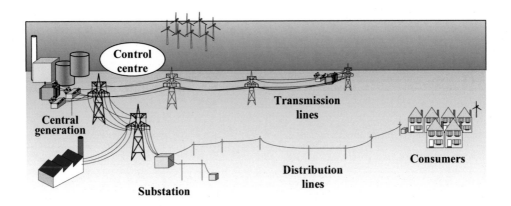

Plate 1 Typical power system elements

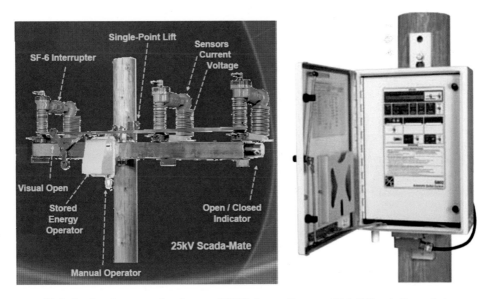

Plate 2 A pole-mounted recloser and RTU. *Source:* Courtesy of S & C Electric Europe Ltd.

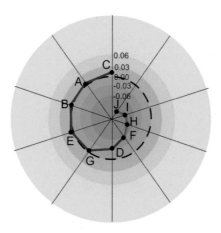

Plate 3 Visual 2-D presentations of nodal voltages. *Source:* Courtesy of Tianda Qiushi Power New Technology Co., Ltd, China

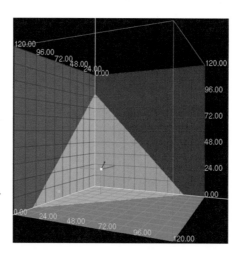

Plate 4 Visual 3-D presentations of a cut-set voltage stability region. *Source:* Courtesy of Tianda Qiushi Power New Technology Co., Ltd, China

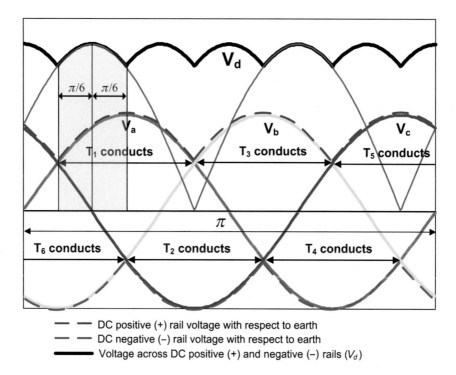

Plate 5 Operation of the CSC without firing angle delay

Plate 6 A TSC–TCR arrangement. *Source:* Photo courtesy of Toshiba

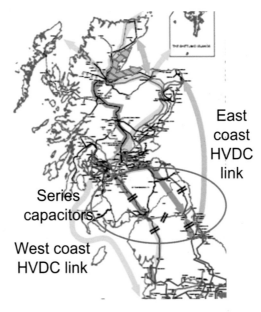

Plate 7 Proposed reinforcements to the UK Transmission network

Plate 8 Thyristor valve. *Source:* Courtesy of Toshiba

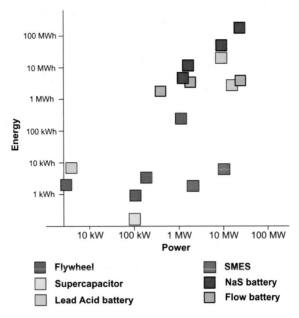

Plate 9   Implemented energy storage schemes based on technologies

Plate 10   Rokkasho wind farm, Japan. *Source:* Courtesy of Japan Wind Development Co. Ltd.

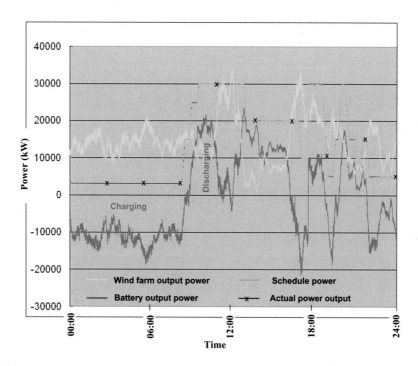

Plate 11 Scheduling wind power with energy storage. *Source:* Courtesy of Japan Wind Development Co. Ltd. *Note* that actual power output and scheduled power output coincide

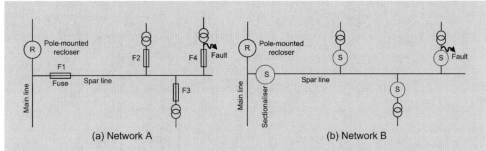

Figure 6.22   Two networks for Example 6.2

Answer

For network A:

- When the fault current is 1000 A (left-hand line of Figure 6.23), the upstream recloser will operate on its instantaneous trip setting. Then after a delay, the recloser is reclosed. If the fault is temporary, the supply will be restored to all the loads.
- When the fault current is 2000 A (right-hand line in Figure 6.23), fuse F4 will operate before the recloser, thus all the consumers in that sub-feeder, lose supply even though the fault is temporary.

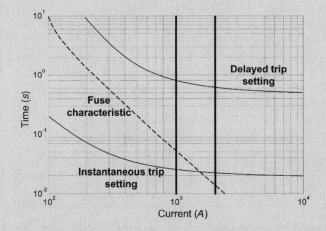

Figure 6.23   Characteristic of fuse F4 and pole-mounted recloser

For network B:

- As the sectionaliser has a pick-up of 200 A, its counter will start immediately after the recloser opens the circuit. As the fault is temporary, the recloser will close the circuit at its subsequent recloses, thus providing power to all the consumers.
- The sectionaliser will reset its counter immediately after the power is restored.

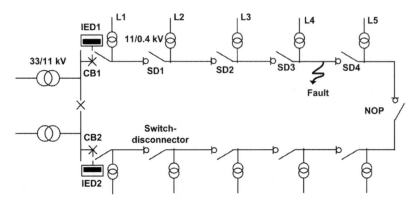

Figure 6.24   Typical distribution network section

6.3.2   Fault location, isolation and restoration

Figure 6.24 shows a typical 11 kV distribution network. When there is a fault on the network at the location shown, the over-current protection element in IED1 detects the fault and opens CB1. This will result in an outage at loads L1 to L5. Since there are no automated components in the network, supply restoration for a part of the network requires the intervention of a restoration crew and in some areas may take up to 80 minutes [20].

Supply restoration is normally initiated by phone calls from one or more customers (in the area where outage occurred) reporting a loss of supply to the electricity supplier. Upon receiving these calls a restoration crew is dispatched to the area. It will take some time for the team to locate the fault and manually isolate it by opening SD3 and SD4. Then CB1 is closed to restore the supply to L1, L2 and L3. The normally open point (NOP) is closed to restore the supply to L5. Load L4 will be without supply until the fault is repaired.

A simple method to reduce the restoration time of loads L1, L2, L3 and L4 is using a pole-mounted recloser and sectionaliser as shown in Figure 6.25. When a fault occurs, the recloser trips. Upon detecting the interruption, the sectionaliser, S, increments its counter by 1.

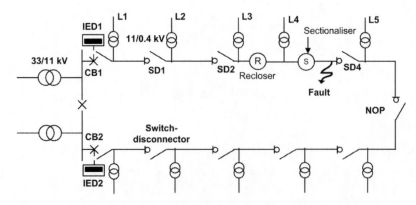

Figure 6.25   Distribution network with some degree of automation

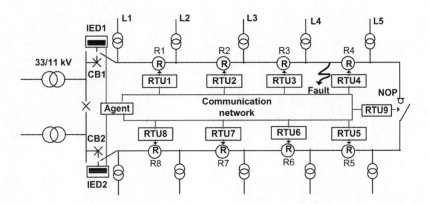

Figure 6.26 Fully automated distribution network

After a short time delay, the recloser closes and if the fault persists, it will trip again. The counter of S increments again and it is then opened. The recloser then closes successfully. The operation of the sectionaliser facilitates restoration of supply to L1, L2, L3 and L4 within a couple of minutes. However, the restoration of supply to L5 requires the intervention of the crew. As this method does not need any communication infrastructure, it is reliable and relatively inexpensive.

A greater degree of automation may be introduced by using reclosers with RTUs, with communication infrastructure between them (see Figure 6.26). In this scheme, an Agent is employed that gathers data from all the intelligent devices in the system. During normal operation, the Agent polls all the RTUs and IEDs to establish the system status. When there is a fault at the location shown, IED1 detects the fault current, opens the CB and informs the Agent. The Agent sends commands to RTU1 to RTU4 (remote terminal units up to the normally open point) to open them and requests current and voltage data from them in real time. A possible automatic restoration method is:

1. Send a command to IED1 to close CB1.
2. Send a command to RTU1 to reclose R1. If the fault current prevails, initiate a trip but as there is no fault current, R1 remains closed. Similarly send commands to RTU2, 3 and 4 to reclose R2, R3 and R4. When R3 is closed, fault current flows, thus causing R3 to trip and lock-out.
3. Then send a command to RTU9 to close the normally open point.
4. Finally, send a command to RTU4 to close R4. As the fault current flows, a trip command is initiated for R4. R3 and R4 thus isolate the fault and supply is restored to loads L1, L2, L3 and L5.

Example 6.3

A section of a distribution network is shown in Figure 6.27. Daily load profiles of each load are given in Figure 6.28. All loads are assumed to be unity power factor:

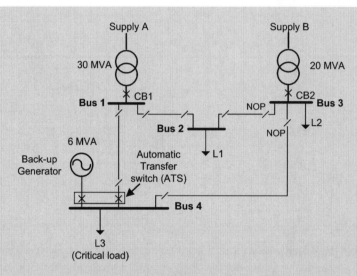

Figure 6.27   Network for Example 6.3

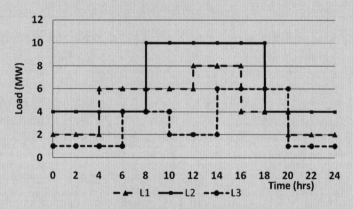

Figure 6.28   Daily load profiles

1. Discuss the consequence of loss of Supply A when there is no automation.
2. Discuss a possible automatic restoration scheme which employs an Agent and re-closers with remote terminal units that provide minimum interruption to all the loads.

Answer

1. *Without any automation*: In the case of the passive network (no automation), the loss of Supply A resulted in automatic transfer of load L3 to the back-up generator. Load L1 will experience an outage. Upon informing the network operator of the outage, a restoration team will be sent to the site. The team, who has access to Figure 6.27

will look at the ratings of different loads and decide on the restoration scheme. As load L3 contains critical loads and as the back-up generator cannot run continuously, the only possible restoration scheme for this network is to close the normally open isolator between buses 3 and 4 and restore the supply to load L3. Load L1 will be without supply until Supply A is restored.

2. *With automatic restoration system*: Figure 6.29 shows the total load on the network. It can clearly be seen that the only time that the total load exceeds the capacity of Supply B is from 14.00–16.00 hrs. However, this is less than the total capacity of Supply B and the output of the back-up generator.

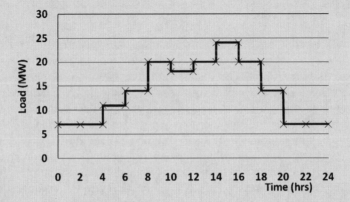

Figure 6.29 Total load on the network

When an Agent is used, it interrogates a database (where load profiles are stored) and stores the load profile of the network in its own memory (which will be refreshed every time the Agent interrogates the database). Immediately after the loss of Supply A, the Agent checks the load profile to decide the possible restoration scheme. For example if the fault occurs at 10.00 hrs, the Agent sends commands to close two NOPs and supply loads L1 and L3 from Supply B with minimum delay (as the total load can be supplied from Supply B). Just before 14.00 hrs a signal is sent to start the back-up generator. At 14.00 hrs, a signal is sent to the ATS and transfers L3 to the back-up generator, thus maintaining supply to loads L1 and L3. At 16.00 hrs, the back-up generator is switched off, connecting load L3 to Supply B. The network configuration is returned to its original configuration immediately after Supply A is restored.

6.4   Voltage regulation

Distribution circuits are subject to voltage variations due to the continuous changes of the network load. At times of heavy load the voltage of the downstream networks is reduced and may go below the lower limit (in the UK the voltage on the 230/400 V circuits should be maintained within +6 per cent and −10 per cent). Under light load conditions the voltage

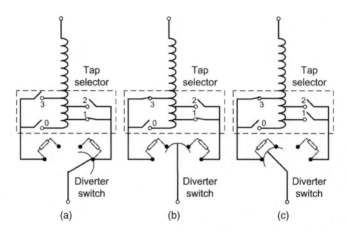

Figure 6.30   Operating sequence of an OLTC

may go above the upper limit. The voltage variations may become severe when distributed generators are connected under light load conditions, the power flow may be reversed [1]. As discussed in Section 6.3.1, sustained voltages above 10 per cent or below −10 per cent of the nominal voltage may damage or may prevent normal operation of IT equipment. As many consumers (domestic, industrial and commercial) are now heavily dependent on this equipment, regulating the voltage within the national limits is very important.

Traditionally, an automatic tap changer and Automatic Voltage Control (AVC) relay, some-times with line drop compensation [21], is used on the HV/MV transformers to maintain the voltages on distribution circuits within limits. These transformers whose output voltage can be tapped while passing load current are referred to as having On-Load Tap Changers (OLTCs). The operation of the OLTC is achieved by operating a tap selector in coordination with a diverter switch as shown in Figure 6.30 a, b, c.

In automatic arrangements (with a motorised tap changer), an AVC relay is introduced to maintain the MV busbar voltage within an upper and lower bounds (a set value ± a tolerance). The main purpose of introducing a tolerance is to prevent the continuous tapping up and down (hunting effect). A time-delay relay is also usually employed to prevent tap changing due to short-term voltage variations. In modern automatic on-load tap changers, the AVC software and time-delay relay are in a bay controller as shown in Figure 6.31.

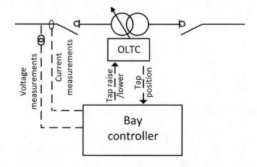

Figure 6.31   Automatic OLTC arrangement

In the USA, pole-mounted capacitor banks are used in distribution circuits for voltage regulation. They provide power factor correction closer to load, improve voltage stability, increase the line power, flows are lower and reduce network losses. These capacitors may be fixed or variable. Modern pole-mounted variable capacitors come with current- and voltage-sensing devices, data logging facility and local intelligence. Variable capacitors are essentially a number of switched capacitors where the number of capacitors that are switched in is determined by an intelligent controller fixed to the pole.

The location of the capacitor bank that provides reactive power and its value of reactive power support are critical to achieve the optimum voltage profile while minimising network losses and maximising line flows. In some cases coordinated control of the OLTC transformer, pole-mounted capacitors and any distributed generators in the network may provide enhanced optimisation of different parameters.

Example 6.4

A section of a distribution network is shown in Figure 6.32.

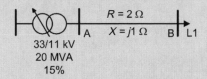

33/11 kV
20 MVA
15%

Figure 6.32 Figure for Example 6.4

1. Calculate the voltage the OLTC should maintain at busbar A to achieve a line voltage of 11 kV at busbar B, when (i) L1 = 1 MW and (ii) L1 = 3 MW.
2. If the 33/11 kV transformer has taps of ±12.5 per cent in steps of 2.5 per cent, discuss the tap setting for cases (i) and (ii). Assume that the primary voltage of the transformer is 33 kV.
3. With the calculated voltage under (a)(i) on busbar A, calculate the value of the capacitor in Mvar required to set the magnitude of the line voltage at busbar B to 11 kV, when L1 = 3 MW.

Answer

Select $S_{base} = 100$ MVA and $V_{base} = 11$ kV

$$Z_{base} = \frac{(V_{base})^2}{S_{base}} = \frac{(11 \times 10^3)^2}{100 \times 10^6} = 1.21 \ \Omega$$

Line impedance in pu $= [2 + j1]/1.21 = 1.65 + j0.83$ pu
Transformer reactance $= 15\%$ on 20 MVA
$$= 0.15 \times 100/20 = 0.75 \text{ pu}$$

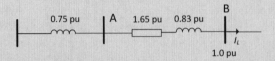

Figure 6.33   pu equivalent circuit

1. (i) When L1 = 1 MW

   As voltage at busbar B is $1\angle0°$ pu, $I_L = S$ in pu $= 1/100 = 0.01$ pu

   Voltage at busbar A $= 1.0 + (1.65 + j0.83)*0.01$

$$= 1.0165 + j0.0083 \text{ pu}$$
$$= 1.0165\angle0.47° \text{ pu}$$

   (ii) When L1 = 3 MW

$$I_L = 0.03 \text{ pu}$$

   Voltage at busbar A $= 1.0 + (1.65 + j0.83)*0.03$

$$= 1.0495 + j0.0249 \text{ pu}$$
$$= 1.05\angle1.36° \text{ pu}$$

2. On a 33/11 kV transformer a tap changer is normally on the 33 kV side. When the
   tap setting is at 0 per cent, the voltage at busbar is 1 pu. Lowering the tap position
   by one step, the voltage becomes 1.025 pu. So under case (a)(i), the tap setting will
   remain at 0 per cent. When the load is increased to 3 MW, in order to maintain the
   required voltage, the tap position should be lowered by two steps, that is, to −5 per
   cent. This gives a voltage of 1.05 pu.
3. With a capacitor connected to busbar B, see Figure 6.34.

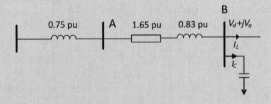

Figure 6.34   Equivalent circuit with a capacitor connected to busbar B

When L1 = 3 MW, $I_L = 0.03$ pu

Assume capacitor current is I_C and as it leads the voltage ($V_d + jV_q$), the current in the
line section AB $= I_L + jI_C$

   Voltage at busbar B $= 1.0165\angle0.47° + (1.65 + j0.83) * (0.03 + jI_C)$

Equating the voltage at busbar B to $V_d + jV_q$:

$$1.0165\angle 0.47° + (1.65 + j0.83) * (0.03 + jI_C) = V_d + jV_q$$
$$[1.0165 + 0.0495 - 0.83I_C] + j[0.0083 + 0.0249 + 1.65I_C] = V_d + jV_q$$

Equating real and reactive parts of the above equation:

$$V_d = 1.066 - 0.83I_C$$
$$V_q = 0.0332 + 1.65I_C$$

As the magnitude of the voltage at busbar B should be 1 pu:

$$[1.066 - 0.83I_C]^2 + [0.0332 + 1.65I_C]^2 = 1$$

Solving this quadratic equation, one can obtain:

$$I_C = 0.106 \text{ pu and } 0.381 \text{ pu}$$

The first solution gives the voltage at busbar B as $1\angle 12°$; whereas the second solution gives it as $1\angle 41.5°$. As the phase angle of the second solution is not reasonable, the first solution was selected.

The value of capacitance required = 1.06 * 10 Mvar = 10.6 Mvar.

It is worth noting that this capacitor is much larger than the load connected at busbar B.

References

[1] Jenkins, N., Ekanayake, J.B. and Strbac, G. (2010) *Distributed Generation*, IET, London.
[2] Djapic, P., Ramsay, C., Pudjianto, D. *et al.* (2007) Taking an active approach. *IEEE Power and Energy Magazine*, 5(4), 68–77.
[3] *Network Protection and Automation Guide: Protective Relays, Measurements and Control*, May 2011, Alstom Grid, Available from http://www.alstom.com/grid/NPAG/ on request.
[4] IEEE Power Engineering Society (2004) *Substation Automation Tutorial*, Document No: 03TP166, March.
[5] Brand, K.P., Brunner, C. and Wimmer, W. (2004) *Design of IEC 61850 Based Substation Automation Systems According to Customer Requirements*, CIGRE session 2004.
[6] Weedy, B.M. and Cory, B.J. (2004) *Electric Power Systems*, John Wiley & Sons, Inc., Hoboken, NJ.
[7] Anderson, P.M. (1999) *Power System Protection*, IEEE and McGraw-Hill, New York.
[8] American National Standard Institute (2008) *ANSI/IEEE C57.13: IEEE Standard Requirements for Instrument Transformers*.
[9] *IEC 60044 (EN 60044 and BS60044): Instrument Transformers - Part 1: Current Transformers, and Part 2: Inductive Voltage Transformers*, 2003.
[10] Blackburn, J.L. and Domin, T.J. (2007) *Protective Relaying: Principles and Applications*, CRC Press, New York.

[11] Areva T&D, *Non-Conventional Instrument Transformer Solutions*, http://www.alstom. com/assetmanagement/DownloadAsset.aspx?ID=80eed284-1e5f-4be0-96ef-8efc0cc 4999c&version=bdffedb3bbb74fb08544f352255ca5321.pdf (accessed on 4 August 2011).

[12] *Optical CTs and VTs*, NXT_{PHASE}, http://nxtphaseinc.com/pdfs/NxtPhase_Optical_ Instrument_Transformers.pdf (accessed on 4 August 2011).

[13] Dugan, R., Santoso, S., McGranaghan, M. and Beaty, H. (2002) *Electric Power Systems Quality*, McGraw-Hill, New York.

[14] Vilathgamuwa, D.M., Wijekoon, H.M. and Choi, S.S. (2006) A novel technique to compensate voltage sags in multiline distribution system: the interline dynamic voltage restorer. *IEEE Transactions on Industrial Electronics*, **53**(5), 1603–1611.

[15] *The ITI (CBEMA) Curve*, Information Technology Industry Council (ITI), http://www.itic.org/clientuploads/Oct2000Curve.pdf (accessed on 4 August 2011).

[16] Arulampalum, A., Ekanayake, J.B. and Jenkins, N. (2003) Application study of a STAT-COM with energy storage. *IEE Proceedings Generation, Transmission and Distribution*, **150**(3), 373–384.

[17] Northcote-Green, J. and Wilson, R. (2007) *Control and Automation of Electrical Power Distribution Systems*, CRC Press, New York.

[18] Stewart, S. (2004) *Distribution Switchgear*, IEE Power Engineering Series 46.

[19] Hart, D.G., Uy, D., Northcote-Green, J., LaPlace, C. and Novosel, D. (2000) Automated solutions for distribution feeders. *IEEE Computer Applications in Power*, **13**(4), 25–30.

[20] Staszesky, D.M., Craig, D. and Befus, C. (2005) Advanced feeder automation is here. *IEEE Power Engineering Magazine*, **3**(5), 56–63.

[21] Lakervi, E. and Holmes, E.J. (1996) *Electrical Distribution Network Design*, IEE Power Engineering Series 21.

7

Distribution Management Systems

7.1 Introduction

Electricity distribution networks connect the high-voltage transmission system to users. Conventional distribution networks have been developed over the past 70 years to accept bulk power from the transmission system and distribute it to customers; generally they have unidirectional power flows. The Smart Grid is a radical reappraisal of the function of distribution networks to include:

- integration of Distributed Energy Resources;
- active control of load demand;
- more effective use of distribution network assets.

Distribution systems are extensive and complex and so they are difficult to monitor, control, analyse and manage. Table 7.1 shows some of the factors that contribute to the complexity of distribution systems.

Real-time monitoring and remote control are very limited in today's electricity distribution systems and so there is a need for intervention by the system operators particularly during widespread faults and system emergencies. However, it is difficult to deal with such a complex system through manual procedures.

A Distribution Management System (DMS) is a collection of Applications used by the Distribution Network Operators (DNO) to monitor, control and optimise the performance of the distribution system and is an attempt to manage its complexity. The ultimate goal of a DMS is to enable a smart, self-healing distribution system and to provide improvements in: supply reliability and quality, efficiency and effectiveness of system operation. A DMS should lead to better asset management, the provision of new services and greater customer satisfaction.

The first generation of Distribution Management Systems integrated a number of simple Applications into a computer system. An interactive graphical user interface was then added to visualise the network being managed. The subsequent use of large relational databases allowed the management of more complicated distribution networks and a large volume

Smart Grid: Technology and Applications, First Edition.
Janaka Ekanayake, Kithsiri Liyanage, Jianzhong Wu, Akihiko Yokoyama and Nick Jenkins.
© 2012 John Wiley & Sons, Ltd. Published 2012 by John Wiley & Sons, Ltd.

Table 7.1 Complexity of distribution systems

Complexity	Source of complexity
Network	Distribution networks are often built as meshed circuits but operated radially. Their topology changes frequently during operation, due to faults and maintenance.
	The structure of the network changes as the network expands
	The three-phases are often unbalanced
	The time scales that have to be considered range from milliseconds (protection operation) to years (network expansion)
	The networks have strict performance objectives
	There is limited communication between elements of the network and most control is local
	Comprehensive monitoring of distribution networks would generate a very large amount of data
Loads	The composition of loads is complex and not well known
	The pattern of distribution load consumption varies dynamically with time. The trend of the load variation is more difficult to predict than that of a large transmission network
	It is not possible to obtain simultaneous measurements of all loads. Load measurements normally are insufficient and may contain large errors and bad data
	The correlation between loads is not well understood

of data. However, as more and more Applications were added, managing the information exchange and maintaining the DMS became a challenge. Standardised models such as the Common Information Model (CIM) were developed to aid information management. For the Smart Grid future, the DMS will use higher-performance ICT hardware, be equipped with greater intelligence, and be deployed in a decentralised manner.

A DMS includes a number of Applications that use modelling and analysis tools together with data sources and interfaces to external systems, as shown in Figure 7.1. The modelling and analysis tools are pieces of software which support one or more Applications.

7.2   Data sources and associated external systems

As shown in Figure 7.1, a DMS includes Applications:

1. for system monitoring, operation and outage management. These are the Applications responsible for the daily running of the network with the primary object of maintaining continuity of supply.
2. to help manage the assets of the utility, such as inventory control, construction, plant records, drawings, and mapping. These include the automated mapping system, the facilities management system, and the geographical information system.
3. associated with design and planning for network extensions. These Applications are used for audits of system operation to determine short-term solutions and optimal expansion planning to achieve system reinforcement at minimum cost.

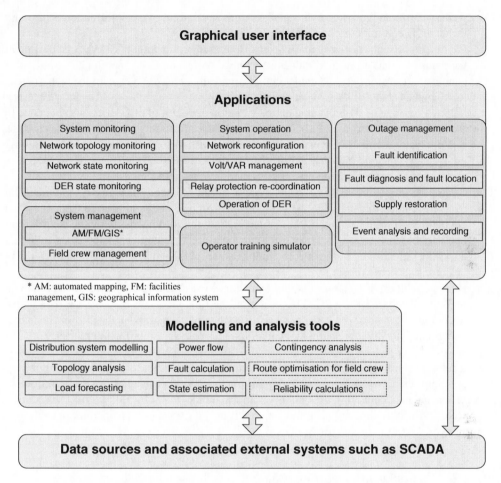

Figure 7.1 Structure and main components of a DMS

All these Applications require modelling and analysis tools for which network parameters, customer information, and network status data are used as inputs.

7.2.1 SCADA

SCADA (Supervisory Control And Data Acquisition) provides real-time system information to the modelling and analysis tools. Hence the data integrity and expandability of the SCADA database are critical. Data integrity should be independent of any DMS Applications and new functions should be able to be integrated easily with the SCADA system without affecting existing Applications.

SCADA has the following attributes:

1. *Data acquisition*: Information describing the system operating state is collected automatically by Remote Terminal Units (RTUs). This includes the status of switching devices as

well as alarms and measured values of voltages and currents. This information is passed to the control centre in close to real-time.

2. *Monitoring, event processing and alarms*; An important function of SCADA is to compare the measured data to normal values and limits, for example, to monitor the overload of equipment (transformers and feeder circuits), and violations of voltage limits. It also detects the change of status of switchgear and operation of protection relays. An event is generated if there is change of switchgear status or violation of circuit limits. All events generated by the monitoring function are processed by the event processing function, which classifies and groups events and delivers appropriate information to the system operators through the Human–Machine Interface (HMI). Most critical events will be sent to the operators as alarms, for example, flashing colour presentation or audible signals.

3. *Control*: Control through a SCADA system can be initiated manually or automatically. Control initiated manually can be the direct control of a particular device (for example, a circuit breaker or tap-changer). Some functions are initiated manually by the control room operator, but then follow local control logic to ensure the equipment is operated following a specific sequence or within specific limits. Control initiated automatically is triggered by an event or specific time.

4. *Data storage, event log, analysis and reporting*: Real-time measurements are stored in the real-time database of the SCADA system at the time received. Because the data update overwrites old values with new ones, the time-tagged data is stored in the historical database at periodic intervals, for example, every 5 minutes or every hour, for future use.

In order to analyse system disturbances correctly, an accurate time-stamped event log is necessary. Some equipment (for example, RTUs) is capable of recording events with millisecond precision and then delivering time-stamped information to the SCADA system. The sequence of events formed by time-stamped information is useful for the system operator to analyse an event to establish the reason for its occurrence.

7.2.2   Customer information system

A Customer Information System (CIS) maintains databases of customers' names, addresses, and network connection. Typical CIS Applications and associated tools are shown in Figure 7.2.

7.3   Modelling and analysis tools

7.3.1   Distribution system modelling

Figure 7.3 shows a typical distribution system. It consists of a number of Medium and Low Voltage feeders that are operated radially. Each Medium to Low Voltage transformer is protected by a circuit breaker or fuse-switch.

7.3.1.1   Single-phase models for balanced three-phase distribution systems

Distribution circuits
A distribution circuit (overhead line or underground cable) can be modelled as a series impedance ($\mathbf{Z}$) with shunt capacitive reactance. The series impedance is composed of

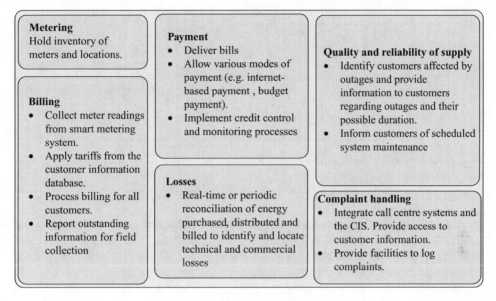

Metering
Hold inventory of
meters and locations.

Billing
- Collect meter readings
  from smart metering
  system.
- Apply tariffs from the
  customer information
  database.
- Process billing for all
  customers.
- Report outstanding
  information for field
  collection

Payment
- Deliver bills
- Allow various modes of
  payment (e.g. internet-
  based payment , budget
  payment).
- Implement credit control
  and monitoring processes

Losses
- Real-time or periodic
  reconciliation of energy
  purchased, distributed and
  billed to identify and locate
  technical and commercial
  losses

Quality and reliability of supply
- Identify customers affected by
  outages and provide
  information to customers
  regarding outages and their
  possible duration.
- Inform customers of scheduled
  system maintenance

Complaint handling
- Integrate call centre systems and
  the CIS. Provide access to
  customer information.
- Provide facilities to log
  complaints.

Figure 7.2   Typical Applications of CIS

resistance (R) and inductive reactance (X_L). The shunt capacitive reactance (X_C) is due to a distribution circuit's natural capacitance. X_C can often be ignored for lower voltage distribution load flow calculations (circuits less than 20 kV) and for distribution fault calculations. Figure 7.4 illustrates a two-port single-phase π-model of a distribution circuit.

Transformers
The basic principles of transformers are covered in standard textbooks [1, 2]. Many distribution transformers are equipped with taps in one winding to change the turns ratio in order to control the system voltage. There are two common types of tap-changers used in distribution transformers: off-circuit tap changers (used in 11/0.4 kV transformers in the UK) and On-Load Tap Changers (OLTC) (used in 33/11 kV transformers). As the names imply, off-circuit tap

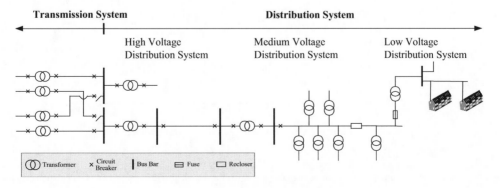

Figure 7.3   A distribution network

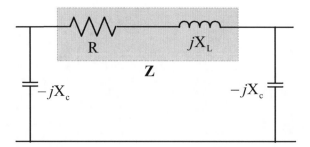

Figure 7.4   Two-port single-phase π-model of a distribution circuit. *Note: J* is the complex operator, square root sign-1

changers require the transformer to be de-energised before the taps are changed while an OLTC can operate with load current continuing to flow through the transformer. Off-circuit taps are set only once, when the transformer is installed, while an OLTC is used to control the system voltage as it varies during the day.

The equivalent circuit for a transformer, Figure 7.5, is widely used to represent two winding transformers. In this model, n is the turns ratio, $\mathbf{Z}_T$ is the transformer impedance representing the winding resistances and leakage reactances, and the subscripts p and s refer to primary and secondary quantities. Magnetising reactance and magnetic losses are ignored.

If the base voltages of the primary and secondary side are related by the turns ratio, the per unit[1] equivalent circuit of a transformer can be represented without the ideal transformer [1, 2]. However, if the base quantities cannot be chosen based on the actual turns ratio or the tap changer needs to be modelled, the equivalent circuit of Figure 7.6 may be used. In this model $\mathbf{Y}_T$ is the admittance of the transformer which is equal to $1/\mathbf{Z}_T$.

Capacitors and reactors
Shunt capacitors can be used for voltage/reactive power control. Series-connected reactors can be used to reduce the fault current. Capacitors and reactors can be modelled using their per phase reactance. The sign of the complex reactance value is determined by the type of the elements (positive for reactors and negative for capacitors).

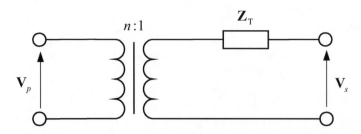

Figure 7.5   Equivalent circuit for a transformer

---

[1] The per unit system is used by power engineers to normalise system quantities and simplify calculations [1, 2].

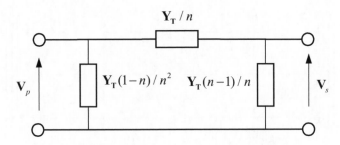

Figure 7.6   Equivalent π-model of a transformer

Loads

Most loads are voltage dependent, which can be represented by an exponential model:

$$P_{Li} = P_{Li\_0}V_i^p$$
$$Q_{Li} = Q_{Li\_0}V_i^q \tag{7.1}$$

where

P_{Li} and Q_{Li} are active and reactive load power at busbar i when the nodal voltage at busbar i is V_i;

P_{Li_0} and Q_{Li_0} are active and reactive load power at busbar i when the nodal voltage at busbar i is 1 pu;

The exponents p and q are 0 when the model represents a constant power load; are 1 when the load is a constant current load; and are 2 when the load is a constant impedance load.

For a composite load, p and q are determined based on the aggregate characteristics of load components.

For steady-state analysis of a distribution network, loads are normally modelled as equivalent complex power absorbers.

Distributed generators

There are two types of nodes in conventional distribution load flow analysis: V-θ nodes and P-Q nodes. The sending end of feeders which is located in a substation is usually modelled as a V-θ node, and all other nodes are modelled as P-Q nodes. When a distribution network has more than one sending end feeder, a strong node is defined as a slack bus (which is always one node and specified as a voltage of constant magnitude and angle).

Distributed Generation (DG) is connected to distribution networks either directly or through power electronic converters. There are four DG grid-coupling techniques which are widely used: directly connected synchronous generator, directly connected induction generator, doubly fed induction generator and full power converters [3]. Some large synchronous generators and voltage-controlled power converters can be modelled as P-V nodes although smaller units often have their control systems set so they operate as P-Q nodes. Induction generators can be modelled as P-Q nodes. For most steady-state studies, simple models are sufficient; however, dynamic studies normally need more detailed machine models.

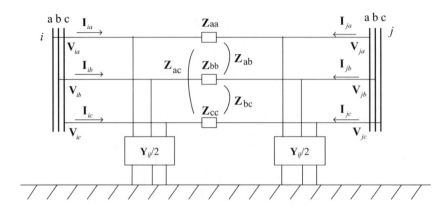

Figure 7.7   Three-phase model of a distribution line

LV and MV networks

LV and MV networks can be modelled using Graph Theory. The terminal nodes of the component models are joined together if physical connections exist. If one of the terminal nodes of a branch is connected to the neutral, it is defined as a shunt branch; branches with all terminal nodes ungrounded are called series branches. Each branch has associated parameters, for example, open/closed status or impedance. Comprehensive analysis can be carried out based on such a graph.

7.3.1.2   Three-phase models for unbalanced three-phase distribution systems

The models discussed so far are single-phase representations of a three-phase system where all the generation and load is balanced across the phases. More detailed modelling of a three-phase system is needed if the system is unbalanced or a large amount of single phase DG or load is connected. Figure 7.7 gives an example of the detailed three-phase modelling of a distribution circuit.

The bus impedance matrix is shown in Equation (7.2) and the shunt admittance matrix of this distribution line is shown in Equation (7.3).

$$\mathbf{Z}_{ij} = \begin{bmatrix} \mathbf{Z}_{aa} & \mathbf{Z}_{ab} & \mathbf{Z}_{ac} \\ \mathbf{Z}_{ba} & \mathbf{Z}_{bb} & \mathbf{Z}_{bc} \\ \mathbf{Z}_{ca} & \mathbf{Z}_{cb} & \mathbf{Z}_{cc} \end{bmatrix} \tag{7.2}$$

$$\mathbf{Y}_{ij}/2 = \frac{1}{2} \times \begin{bmatrix} \mathbf{Y}_{aa} & \mathbf{Y}_{ab} & \mathbf{Y}_{ac} \\ \mathbf{Y}_{ba} & \mathbf{Y}_{bb} & \mathbf{Y}_{bc} \\ \mathbf{Y}_{ca} & \mathbf{Y}_{cb} & \mathbf{Y}_{cc} \end{bmatrix} \tag{7.3}$$

The detailed three-phase models of distribution system equipment are much more complicated than the single-phase model. For example, three-phase transformer models consider not only the winding resistances and leakage reactance, but also the methods of three-phase connection. Delta, Grounded Wye (or Star). A different winding arrangement will give a different transformer admittance matrix.

More details of three-phase distribution modelling can be found in [4].

7.3.1.3 New trends for smart grids

At present the parameters of distribution system models are obtained from manufacturers' data, historical information, or site tests. With changes of external conditions, for example, ambient temperature, or the ageing of equipment, such parameters can change over time and introduce errors into the network modelling and hence result in unreliable system operation.

The ICT infrastructure of a Smart Grid provides the opportunity for more accurate system modelling through the application of statistical system identification techniques. System identification, widely used in control engineering, can be used to build mathematical models of a distribution system based on the large amount of data measured by the ICT system. A set of more accurate, continuously updated models can then be obtained, and used by the DMS Applications.

7.3.2 Topology analysis

An electric power distribution network consists of a variety of equipment which must be modelled in a concise and standard form for power systems analysis. The mapping between the physical plant model and the power systems analysis model is undertaken by the topology analysis tool. The topology analysis tool carries out network reduction. This reduces the amount of data feeding into other modelling and analysis tools and allows easier interpretation of results by the operator. For example, a substation that contains six sections of busbar[2] which are linked by several open/closed items of switchgear may be represented by a single electrical node for power system analysis. The topology analysis tool is able to distinguish the energised parts of the power system from the de-energised parts, and can identify the electrically separated 'islands'.

Figure 7.8(a) shows a one-line diagram of a distribution network, which is represented by physical plant model with 54 busbars, and Figure 7.8(b) provides the equivalent power system analysis model with only 13 nodes derived from the topological tool. As can be seen by the encircled section in Figure 7.8(a), a number of circuit breakers/switches/fuses and busbar sections are represented by a single node (A) in the power system analysis model shown in Figure 7.8(b).

Commonly used graph searching algorithms for topology analysis are depth-first and breadth-first methods [5]. A depth-first search starts at the root and explores as far as possible along each branch before backtracking. A breadth-first search begins at the root node and explores all the neighbouring nodes, then for each of those nearest nodes, it explores their unexplored neighbouring nodes, and so on, until it finds the goal.

The requirements for software implementation include being:

1. *Reliable*: The topology analysis tool should be able to deal with all kinds of network topologies and provide an accurate network model. Any mistake in the derived network model can put the power system into a vulnerable condition.
2. *Efficient*: The time complexity of many topology analysis algorithms is $O(N^2)$ (big O notation[3]), where N is the number of nodes. For a distribution network with a large number

---

[2] A busbar is simply a zero (or very low) impedance connector of several circuits.

[3] In computer science, the time complexity of an algorithm quantifies the amount of run time as a function of the size of the input, and it is commonly expressed using 'big O notation'.

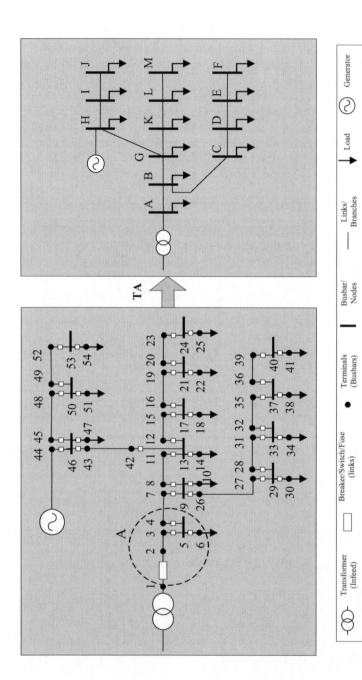

Figure 7.8 Example of topology analysis

Table 7.2 Glossary of terms used in topology analysis

Terms	Description
Infeeds	An infeed represents the power input from an upstream network that is not modeled. The infeed is connected to a busbar in the physical plant model. An energised "island" will be obtained if the infeed status is "ON".
Generators	A generator is connected to a busbar in the physical plant model. An energised "island" will be obtained if the generator status is "ON".
Links	A link is a zero impedance connection between two busbars, e.g. a circuit breaker, an isolator, or a very short cable. Each link has a status ("ON" or "OFF"). A link with "ON" status will merge the two busbars at each end into one node.
Loads	A load is connected to a busbar in the physical plant model.
Nodes	Nodes are outputs of the topology analysis. They represent a set of busbars connected by closed links.
Islands	Islands are outputs of the topology analysis. They represent a set of energised nodes connected by live branches. Electrically separate islands can be identified, and a separate power flow analysis is needed by each island with a separate slack node.
Branches	A branch has non-zero impedance, e.g. an overhead line, a cable, or a transformer. A branch is live if its status is "ON" and is part of an energised "island"; or dead if its status is "OFF" or is part of a de-energised network.

of nodes, the topology analysis will be time-consuming. However, topology analysis is an input to other DMS Applications, and so the performance of the topology analysis tool is important in developing a real-time DMS.

Table 7.2 presents glossary of terms often used in topology analysis.

7.3.3 Load forecasting

There are a very large number of loads in a distribution network and the energy consumption of each load is relatively small. Therefore, comprehensive real-time measurement of loads is too expensive and load estimation and forecasting are used for both the operation and planning of a distribution network.

Load forecasting can be divided into three forecasting time horizons: short-term load forecasting which is usually from one hour to one week; medium-term load forecasting which is usually from one week to one year; and long-term forecasting which is longer than one year. Load forecasting can also be divided into two categories based on the forecasting scope: regional load forecasting which provides load forecasts for a large geographical area and busbar load forecasting which provides nodal load information for network control functions. For a number of DMS Applications, short-term load forecasting is of most significance.

Load varies over the short term with: time (e.g. weekday, weekend, and holiday), weather (e.g. temperature and humidity), and the types of energy consumers (e.g. residential, commercial and industrial). Various methods are available for short-term load forecasting, including:

1. *Regression*: Regression methods are widely used to model the correlation between energy consumption and the influence factors such as time and weather. The forecast value can be represented as

$$F(t) = a + \sum_{i=0}^{N} b_i x_i(t) \qquad (7.4)$$

   where
   $F(t)$ is the forecast load value at time t
   $x_i(t)$ is the ith influence factor at time t
   a, b_i are regression factors
   N is the number of influence factors considered.

2. *Time series*: Time series methods assume that the load data has inherent rules, for example, daily, weekly and seasonal variation, trend, and autocorrelation. Time series methods try to identify and explore such rules. ARMA (Autoregressive Moving Average), ARIMA (Autoregressive Integrated Moving Average), ARMAX (Autoregressive Moving Average with exogenous variables), and ARIMAX (Autoregressive Integrated Moving Average with exogenous variables) are widely used time series methods.

3. *Similar day*: Similar day methods use historical load data of days with similar characteristics to the forecast day. Similar characteristics include day of the week, day of the month, day of the year and weather. In practical applications, the forecast is obtained through a linear combination or regression of several similar days to reduce error.

4. *Computational intelligence-based methods*: These methods include artificial neural networks, knowledge-based engineering, fuzzy logic systems, and evolutionary computation.

5. *Hybrid methods*: Hybrid methods take advantage of several of the above methods and combine them to create a more powerful method.

The development of smart metering and home energy management systems will change the behaviour of energy consumers and lead to more dynamic demand (for example, load that is sensitive to price) and hence loads that are more volatile. More volatile loads are likely to be more difficult to forecast.

7.3.4   Power flow analysis

Power flow (or load flow) analysis provides the steady-state solution of a power network for specific network conditions which include both network topology and load levels. The power flow solution gives the nodal voltages and phase angles and hence the power injections at all buses and power flows through lines, cables and transformers. It is the basic tool for analysis, operation, and planning of distribution networks.

Table 7.3 Types of buses in power flow analysis

Bus type	Quantities specified	Quantities to be obtained
Slack Bus	$|V|$, θ	P, Q
PV Bus	P, $|V|$	Q, θ
PQ Bus	P, Q	$|V|$, θ

In a power system, each busbar is associated with four quantities: the magnitude of voltage ($|V|$) and its angle (θ), real power injection (P) and reactive power injections (Q). For power flow analysis, only two of these quantities are specified, and the remaining two are obtained by the power flow solution. Depending upon the specified and unspecified quantities, the busbars are classified into three types as shown in Table 7.3.

The power flow can be formulated as a set of nonlinear algebraic equations and then a suitable mathematical technique such as the Gauss-Seidel, Newton-Raphson, or fast-decoupled method can be chosen for the solution of the equations. The classical Gauss-Seidel, Newton-Raphson, and fast-decoupled method are described in many textbooks [1, 2, 3].

Although power flow computation for transmission systems is well understood, distribution power flow analysis can encounter difficulties in an ill-conditioned distribution network due to:

- the network structure of a distribution system being radial or weakly meshed;
- the low X/R ratio in a distribution feeder;
- unbalanced loads;
- a mix of short line segments with low impedance and long feeders with high impedance;
- connection of distributed generators.

Therefore, a number of power flow methods have been developed, or adapted from transmission power flow methods, for the analysis of distribution networks. For example, the adapted Newton-Raphson method uses power mismatches at the ends of feeders and laterals to deal with the nodal voltages and can accelerate the convergence of the algorithm. The adapted Gauss-Seidel method uses the bus-impedance matrix to deal with the branch currents.

The forward/backward method was designed specifically for radial or weakly meshed distribution networks [6]. The system infeed bus is chosen as the slack bus, and then the following procedure is used:

1. Assume an initial nodal voltage magnitude and angle for each bus (a nominal voltage is often used).
2. Start from the root and move forward towards the feeder and lateral ends while calculating

$$\mathbf{I}_i^{(k)} = \left[\frac{S_i}{\mathbf{V}_i^{k-1}} \right]^* \tag{7.5}$$

3. Start from the feeder and lateral ends towards the root while calculating

$$\mathbf{V}_i^{k+1} = \mathbf{V}_j^k + \mathbf{Z}_{ij}\mathbf{I}_{ij}^k \tag{7.6}$$

where j is the adjacent down-stream bus to bus i, and two busbars are connected by a branch having impedance of $\mathbf{Z}_{ij}$.

4. Check the termination criterion through calculating the power mismatch (or the voltage mismatch):

$$\Delta S_i^k = S_i - \mathbf{V}_i^k \left[\mathbf{I}_i^k\right]^* \leq \varepsilon \tag{7.7}$$

where ε is the given threshold.

5. The forward/backward sweep is repeated until the criterion set in Equation (7.7) is met.

A weakly meshed network may be converted to a radial network by breaking the loops at a point of low current flow and substituting an equivalent injection current.

Example 7.1

For the network shown in Figure 7.9, show two iterations of the forward/backward method.

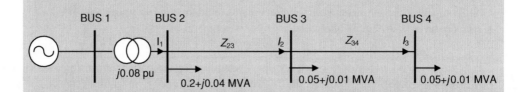

$$Z_{23} = 0.05+j0.07 \text{ pu}$$
$$Z_{34} = 0.05+j0.07 \text{ pu}$$

All quantities are in per unit on 20 kV, 100 MVA base

Figure 7.9   Figure for Example 7.1

Answer

1. *Step 1:* Assume all BUS voltages are $1\angle 0°$ pu; that is,

$$V_1^{(1)} = V_2^{(1)} = V_3^{(1)} = V_4^{(1)} = 1.0$$

2. *Step 2: Forward sweep*
   Current through BUS 1 to 2:

$$I_1^{(1)} = \left[\frac{0.2 + j0.04 + 0.05 + j0.01 + 0.05 + j0.01}{1}\right]^* = 0.3 - j0.06 \text{ pu}$$

   Current through BUS 2 to 3:

$$I_2^{(1)} = \left[\frac{0.05 + j0.01 + 0.05 + j0.01}{1}\right]^* = 0.1 - j0.02 \text{ pu}$$

   Current through BUS 3 to 4:

$$I_3^{(1)} = \left[\frac{0.05 + j0.01}{1}\right]^* = 0.05 - j0.01 \text{ pu}$$

3. *Step 3: Backward sweep*
   Voltage at BUS 3: $V_3^{(2)} = V_4^{(1)} + (0.05 + j0.07) \times I_3^{(1)}$

$$= 1.0 + (0.05 + j0.07) \times (0.05 - j0.01)$$

$$= 1 + 0.086\angle 54.4° \times 0.051\angle -11.3°$$

$$= 1 + 0.004\angle 43.2°$$

$$= 1.003 + j0.003$$

   Voltage at BUS 2: $V_2^{(2)} = V_3^{(1)} + (0.05 + j0.07) \times I_2^{(1)}$

$$= 1.0 + (0.05 + j0.07) \times (0.1 - j0.02)$$

$$= 1 + 0.086\angle 54.4° \times 0.102\angle -11.3°$$

$$= 1 + 0.009\angle 43.2°$$

$$= 1.006 + j0.006 \text{ pu}$$

   Voltage at BUS 1: $V_1^{(2)} = V_2^{(1)} + j0.08 \times I_1^{(1)}$

$$= 1.0 + j0.08 \times (0.3 - j0.06)$$

$$= 1.005 + j0.024 \text{ pu}$$

4. *Step 4: Forward sweep with new voltages*
   Current through BUS 1 to 2:

$$I_1^{(1)} = \left[\frac{0.2 + j0.04 + 0.05 + j0.01 + 0.05 + j0.01}{1.005 + j0.024}\right]^* \text{ pu}$$

$$= \frac{0.3 - j0.06}{1.005 - j0.024} = 0.3043\angle -9.94°$$

$$= 0.2997 - j0.0525 \text{ pu}$$

$$\text{Current through BUS 2 to 3: } I_2^{(1)} = \left[\frac{0.05 + j0.01 + 0.05 + j0.01}{1.006 + j0.006 \, \text{pu}} \right]^* \text{pu}$$

$$= \frac{0.1 - j0.02}{1.006 - j0.006} = 0.1014\angle - 10.97°$$

$$= 0.0995 - j0.0193 \, \text{pu}$$

$$\text{Current through BUS 3 to 4: } I_3^{(1)} = \left[\frac{0.05 + j0.01}{1.003 + j0.003} \right]^* \text{pu}$$

$$= \frac{0.05 - j0.01}{1.003 - j0.003} = 0.0508\angle - 11.14°$$

$$= 0.0498 - j0.0098 \, \text{pu}$$

5. *Step 5: Backward sweep with new currents*
   Voltage at BUS 3:

$$= 1 + 0.086\angle 54.4° \times 0.0508\angle - 11.14°$$
$$= 1 + 0.0044\angle 43.26°$$
$$= 1.0032 + j0.003$$

Voltage at BUS 2: $V_2^{(2)} = V_3^{(1)} + (0.05 + j0.07) \times I_2^{(1)}$

$$= 1 + 0.086\angle 54.4° \times 0.1014\angle - 10.97°$$
$$= 1 + 0.0087\angle 43.43°$$
$$= 1.0063 + j0.006 \, \text{pu}$$

Voltage at BUS 1: $V_1^{(2)} = V_2^{(1)} + j0.08 \times I_1^{(1)}$

$$= 1.0 + j0.08 \times (0.2997 - j0.0525)$$
$$= 1.0042 + j0.024 \, \text{pu}$$

7.3.5　Fault calculations

A distribution network is designed so that it is safe and remains undamaged under both normal and abnormal conditions. The dimensioning, cost effectiveness and safety of these systems (particularly in cities) depend to a great extent on being able to control short circuit currents. With the increasing power of loads and the ratings of distributed generators, the importance of short-circuit current calculation increases. Accurate fault calculations are a prerequisite for the correct dimensioning of electrical equipment, setting of protective devices and ensuring stability. Two types of short-circuit faults are usually considered: symmetrical (balanced) faults and asymmetric (unbalanced) faults.

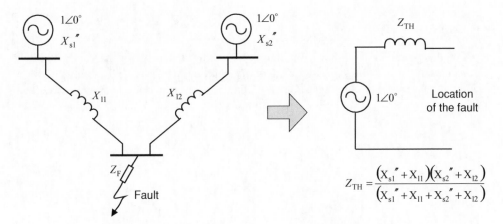

Figure 7.10 Modelling of the network for symmetrical faults

During a fault, currents flow from the network into the ground or into another phase of the network. In order to calculate these currents, a model of the faulted system is required.

7.3.5.1 Symmetrical fault calculation

A symmetrical fault affects all three phases in the same way, therefore the symmetry of the system is retained. The single phase representation of the three-phase system can be used and models of lines, cables, transformers are the same as for normal operating conditions and load flow calculations. Generators are modelled as a constant voltage behind the sub-transient reactance. All parameters are expressed on a common per unit base. For simple fault calculations, as voltage of all the generators are taken as $1\angle 0°$ pu, all parallel connected sources are replaced by the Thevenin equivalent as shown in Figure 7.10.

The fault current is calculated by

$$I_F = \frac{Z_{TH}}{Z_{TH} + Z_F} = \frac{cV_n}{Z_{TH} + Z_F} \tag{7.8}$$

where Z_F is the fault impedance; Z_{TH} is the Thevenin equivalent impedance; $V_n = 1\angle 0°$ is the nominal voltage of the fault location, and c is a voltage factor, (defined in IEC 60909) to adjust the value of the equivalent voltage source for the maximum and minimum fault current conditions. An alternative approach is to determine the pre-fault voltages of the network using a load flow.

When a solid fault occurs ($Z_F = 0$), then the largest fault current results, which is important for sizing the circuit breakers and for stability studies. When $Z_F \neq 0$, then the fault current is reduced and protection should be designed to work with minimum fault current. IEC 60909 provides authoritative guidance on carrying out symmetrical fault calculations.

Example 7.2

For the power system shown in Figure 7.11, draw the network diagram giving all reactances on 100 MVA base. Calculate the fault current in pu and in amperes for a three-phase short-circuit fault at C. All pre-fault voltages are at 1.0 pu. Ignore any effect of system loads.

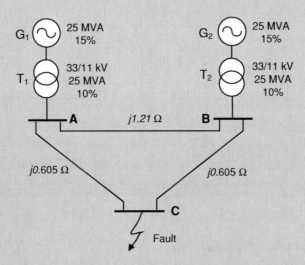

Figure 7.11   Figure for Example 7.2

Answer

1. Base impedance is given by:

$$Z_B = \frac{V_B^2}{S_B} = \frac{(11)^2}{100} = 1.21 \ \Omega$$

Per unit reactance on 100 MVA base	
G_1 0.15 × 100/25 = 0.6 pu	
G_2 0.15 × 100/25 = 0.6 pu	$X_{new} = X_{old} \times \dfrac{S_{new}}{S_{old}}$
T_1 0.1 × 100/25 = 0.4 pu	
T_2 0.1 × 100/25 = 0.4 pu	
Line AB 1.21/1.21 = 1.0 pu	
Line BC 0.605/1.21 = 0.5 pu	$X_{pu} = \dfrac{X}{Z_B}$
Line AC 0.605/1.21 = 0.5 pu	

The per unit equivalent circuit is shown in Figure 7.12.

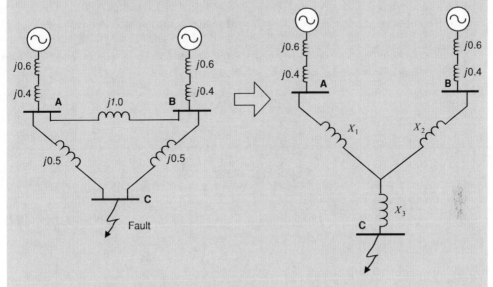

Figure 7.12   pu equivalent circuit of Figure 7.11

To obtain the Thevenin equivalent, it is required to convert the ABC delta loop into an equivalent star [2] as shown in Figure 7.12.
Then

$$X_1 = j\frac{0.5 \times 1}{2} = j0.25$$

$$X_2 = j\frac{0.5 \times 1}{2} = j0.25$$

$$X_3 = j\frac{0.5 \times 0.5}{2} = j0.125$$

Therefore,

$$Z_{\text{TH}} = \frac{j0.6 + j0.4 + j0.25}{2} + j0.125$$

$$= j0.75 \text{ pu}$$

Fault current in per unit $I_F = \dfrac{V_{\text{TH}}}{|Z_{\text{TH}}|} = \dfrac{1.0}{0.75} = 1.33\text{pu}$

$$I_B = \frac{S_B}{\sqrt{3}V_B} = \frac{100 \times 10^6}{\sqrt{3} \times 11 \times 10^3} = 5248.6\text{A}$$

Fault current in A $I_F = 1.33 \times 5248.6 = 6980.7\text{A}$

7.3.5.2 Asymmetrical short circuit fault calculation

Asymmetrical short circuit faults include line-to-line short circuit, line-to-line-to-earth short circuit and line-to-earth short circuit (the most common fault in overhead line circuits). Symmetrical components [1, 2, 3] are used for the calculation of the fault current. Here, the currents flowing in each line are determined by superposing the currents of three symmetrical components: positive sequence current $I_{(1)}$, negative sequence current $I_{(2)}$ and zero sequence current $I_{(0)}$. Therefore, the current I_A, I_B, and I_C can be represented as

$$I_A = I_{(1)} + I_{(2)} + I_{(0)} \tag{7.9}$$

$$I_B = a^2 I_{(1)} + a I_{(2)} + I_{(0)} \tag{7.10}$$

$$I_C = a I_{(1)} + a^2 I_{(2)} + I_{(0)} \tag{7.11}$$

where $\mathbf{a}$ is an operator and given by:

$$\mathbf{a} = -\frac{1}{2} + j\frac{\sqrt{3}}{2} \text{ and } \mathbf{a}^2 = -\frac{1}{2} - j\frac{\sqrt{3}}{2}$$

7.3.6 State estimation

Real-time monitoring of distribution systems is very limited due to the lack of sensors and communication systems. Hence the distribution system can be described as under-determined with the number of measurements insufficient to make the system observable. Once the complete system state is available, then any quantity in the system can be calculated. The observability and controllability of a system are mathematical duals, which means an unobservable system cannot be fully controlled. Grid integration of distributed energy resources (which includes distributed generators, electric vehicles, heat pumps and controllable loads) brings significant uncertainties and, at high penetrations, may lead to operational difficulties in a network. Therefore, the provision of accurate system state information to the network operators is critical for them to operate the system in a safe, prompt, and cost-effective manner, while making the best use of the assets.

State estimation is used to clean up errors in measurements and estimate the system state [7]. State estimation techniques are widely used in transmission systems where redundant measurements are available (that is, the system is over-determined). For under-determined distribution systems, a large number of pseudo measurements from load estimates and load forecasts have to be used as inputs to the distribution state estimator.

Consider the two-busbar system shown in Figure 7.13. The receiving end voltage is given by: $V_2 = V_1 - I_1(R + jX)$. Based on the true value of network parameters, that is, $V_1 = x_1$ and $I_1 = x_2$, the true value of Bus 2 voltage is given by $[x_1 - x_2(R + jX)]$. Assume that the voltage of Bus 2 is measured as z_2. If the error in z_2 is e_2, then the true value of voltage at Bus 2 and measured value of voltage at Bus 2 can be related by Equation (7.12):

$$z_2 = [x_1 - x_2(R + jX)] + e_2 \tag{7.12}$$

In more general form, for a larger network Equation (7.12) could be written as:

$$z = h(x) + e \tag{7.13}$$

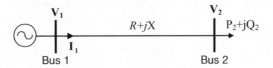

Figure 7.13   Two busbar system

where $\mathbf{z}$ is the measurement matrix, $\mathbf{h}$ is the measurement function governed by circuit equations, $\mathbf{x}$ is the state variable matrix and $\mathbf{e}$ is the measurement error matrix.

The true value of $\mathbf{x}$ cannot be determined, but can be estimated. Assume that the estimated value of $\mathbf{x}$ is $\hat{\mathbf{x}}$ and the estimated value of measurement errors are $\mathbf{r}$ (often this is called the measurement residual), then Equation (7.13) can be rewritten as:

$$\mathbf{z} = \mathbf{h}(\hat{\mathbf{x}}) + \mathbf{r} \tag{7.14}$$

State estimation (determining $\hat{\mathbf{x}}$) can be considered as an optimisation problem, which minimises the measurement residual. This is achieved by minimising a function of the measurement residual, $\rho(r_i)$ as described in Sections 7.3.6.1 and 7.3.6.2. In order to use the measurements from the meters of known greater accuracy with those of lesser accuracy, a measurement weight is introduced to the measurement residual r_i. This can be expressed mathematically as:

$$\min \sum_{i=1}^{m} w_i \rho(r_i) \tag{7.15}$$

where m is the number of measurements and w_i is the weight.

Commonly used state estimation methods include the Weighted Least Square (WLS) method, the Weighted Least Absolute Value (WLAV) method and the Robust state estimation [7].

7.3.6.1   Weighted least square (WLS) estimators

For a WLS estimator, the $\rho(r_i)$ in Equation (7.15) is defined as

$$\rho(r_i) = (z_i - h_i(x))^2 \tag{7.16}$$

Therefore Equation (7.15) can be rewritten as a minimisation problem with objective function $F(\mathbf{x})$

$$\min F(\mathbf{x}) = \min \sum_{i=1}^{m} w_i (z_i - h_i(x))^2 = [\mathbf{z} - \mathbf{h}(\mathbf{x})]^T \mathbf{W}[\mathbf{z} - \mathbf{h}(\mathbf{x})] \tag{7.17}$$

The solution of Equation (7.17) provides the estimated state $\hat{\mathbf{x}}$ that must satisfy the necessary optimality conditions shown in Equation (7.18):

$$\frac{\partial F(\mathbf{x})}{\partial \mathbf{x}} = \mathbf{H}^T(\hat{\mathbf{x}})\mathbf{W}[\mathbf{z} - \mathbf{h}(\mathbf{x})] = 0 \tag{7.18}$$

where $\mathbf{H}$ is the Jacobian matrix of the measurement functions $\mathbf{h}(\mathbf{x})$.

The solution is usually obtained by an iterative method derived by linearising $\mathbf{h}(\mathbf{x})$ around $\mathbf{x}^k$. The iteration uses an initial guess for the system state $\mathbf{x}^0$ as a starting point, and at each iteration k, a set of linear equations, as shown in Equation (7.19). These are called the normal equations of the WLS problem and are solved to calculate the correction $\Delta\mathbf{x}^k$:

$$\mathbf{G}(\mathbf{x}^k)\Delta\mathbf{x}^k = \mathbf{H}^T(\mathbf{x}^k)\mathbf{W}[\mathbf{z} - \mathbf{h}(\mathbf{x}^k)] \tag{7.19}$$

where $\mathbf{G}(\mathbf{x})$ is the gain matrix defined as:

$$\mathbf{G}(\mathbf{x}) = \mathbf{H}^T(\mathbf{x})\mathbf{W}\mathbf{H}(\mathbf{x}) \tag{7.20}$$

The system state is updated using Equation (7.21) until $\Delta\mathbf{x}$ is smaller than a threshold.

$$\mathbf{x}^{k+1} = \mathbf{x}^k + \Delta\mathbf{x}^k \tag{7.21}$$

7.3.6.2   Weighted least absolute value (WLAV) estimators

The Weighted Least Absolute Value (WLAV) estimator is an alternative method for power system state estimation. For a WLAV estimator, the $\rho(r_i)$ in Equation (7.15) is defined as

$$\rho(r_i) = |z_i - h_i(x)| \tag{7.22}$$

Simplex-based algorithms and interior-point-based algorithms have been developed to provide solution algorithms for Equations (7.15) and (7.22).

7.3.6.3   Robust state estimators

Because a large number of pseudo measurements must be used to make the distribution system observable, the distribution state estimator should be robust to the presence of errors and bad data in the pseudo measurements.

For conventional WLS estimators, bad data is detected, identified and removed after the estimation of the system state through processing the measurement residuals, for example, the largest normalised residual test was developed to detect and identify bad data. The measurements will be suspected to be polluted by bad data if any of the normalised residuals is above a given detection threshold. This threshold is defined based on the desired level of confidence that no measurement contains bad data. In addition, hypothesis testing identification techniques have been used for bad data identification to overcome the deficiency of the largest normalised residual test method.

The measurement configuration (type, location, accuracy of measurements) also has a large impact on the quality of the estimates. For some measurements, the residuals ($\mathbf{r}$) may not reflect the measurement errors ($\mathbf{e}$); therefore bad data in such measurements cannot be detected and identified by the measurement residuals (such measurements are called leverage measurements). Robust state estimation techniques have been developed to mitigate the impact from both measurement errors and the measurement configurations. A good design of $\rho(r_i)$ will be able to provide the robustness required. For example, in the iteratively reweighted least square estimation method measurement weights are updated at each iteration.

Example 7.3

For the simple circuit shown in Figure 7.14, the state variable x is the line current I. The measurements are $I = 1.02$ A, $U = 9.99$ V, and $P = 9.95$ W. Calculate the estimates of I, U and P and residuals:

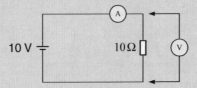

Figure 7.14   Figure for Example 7.3

1. using the least square estimation technique (weighted least square estimation technique with $w_i = 1$ for all i)
2. using the weighted least square estimation technique assuming the measurement errors are I: 0.05 A, U: 0.02 V, P: 0.1 W.

Answer

1. Assume that true value of the current through the circuit is x and the measured quantities of current, voltage and power are z_I, z_U and z_P. The measured values can be related to the true values derived from $V = IR$ and $P = I^2R$ with corresponding errors (e_I, e_U and e_P) as:

$$z_I = x + e_I$$
$$z_U = 10x + e_U$$
$$z_P = 10x^2 + e_P$$

From Equation (7.17), the objective function is defined as:

$$\min F(x) = (1.02 - x)^2 + (9.99 - 10x)^2 + (9.95 - 10x^2)^2$$

$$\frac{\partial F(x)}{\partial x} = -2 \times (1.02 - x) - 20 \times (9.99 - 10x) - 40 \times x \times (9.95 - 10x^2)$$

Making $\dfrac{\partial F(x)}{\partial x} = 0$:

$$200x^3 - 98x - 100.92 = 0$$

$$\therefore x = 0.9978$$

Estimated values

$$I = x = 0.9978\,\text{A}$$
$$U = 10x = 9.978\,\text{V}$$
$$P = 10x^2 = 9.956\,\text{W}$$

Residuals

$$r_I = 1.02 - 0.9978 = 0.0222\,\text{A}$$
$$r_U = 9.99 - 9.978 = 0.012\,\text{V}$$
$$r_P = 9.95 - 9.956 = -0.006\,\text{W}$$

2. The measurement errors are I: 0.05 A, U: 0.02 V, P: 0.1 W. So the corresponding weights can be chosen as:

$$I : 1/0.05^2 = 400,$$
$$U : 1/0.02^2 = 2500,$$
$$P : 1/0.1^2 = 100$$

Define the objective function:

$$\min F(\mathbf{x}) = \min \sum_{i=1}^{m} w_i (z_i - h_i(x))^2$$

$$= 400 \times (1.02 - x)^2 + 2500 \times (9.99 - 10x)^2 + 100 \times (9.95 - 10x^2)^2$$

Making $\dfrac{\partial F(x)}{\partial x} = 0$:

$$\frac{\partial F(x)}{\partial x} = -800 \times (1.02 - x) - 50000 \times (9.99 - 10x) - 4000 \times x \times (9.95 - 10x^2)$$

$$= 0$$

$$x^3 + 11.525x - 12.5079 = 0$$

$$x = 0.9988$$

Estimated values:

$$I = x = 0.9988\,\text{A}$$
$$U = 10x = 9.988\,\text{V}$$
$$P = 10x^2 = 9.976\,\text{W}$$

Residuals:

$$r_I = 1.02 - 0.9988 = 0.0212\,\text{A}$$
$$r_U = 9.99 - 9.988 = 0.002\,\text{V}$$
$$r_P = 9.95 - 9.976 = -0.026\,\text{W}$$

7.3.7   Other analysis tools

7.3.7.1   Contingency analysis

Contingency analysis is widely used in transmission system management. However, it is being extended to the distribution system with the development of smart distribution networks. This tool performs N-K contingency analysis in a distribution network (N-1 and N-2 are normally used), which informs the network operators of the vulnerability of the distribution system in real time. It is mainly used for operation planning of a distribution network. For a set of contingencies (possible faults), the contingency analysis tool returns the ranked severity of all the contingencies considered and for each contingency, the tool also returns the optimum remedial actions.

This tool can be run periodically, triggered by events (topology changes, loading condition changes, control availability), or operated in study mode for operator training.

The factors which need to be considered in distribution contingency analysis for the Smart Grid include the impact of Distributed Energy Resources (DERs) and MicroGrids/Cells (see Section 7.4.2.4).

7.3.7.2   Route optimisation for field crews

Route optimisation is an analysis tool for the supervision of the field maintenance crews and optimisation of their driving routes. The road data and terrain conditions of the distribution networks are stored in the geographical database. Transport models are integrated into the tool to represent the travelling speed and transition time, which will be used to optimise the routes and calculate the transit time of the crew.

7.4   Applications

The automation of operation and management of distribution network requires a large number of Applications.

7.4.1   System monitoring

Real-time distribution system monitoring can bring a number of benefits to system operation. For example, it can lead to better information of nodal voltages and circuit loading conditions, which allows alarms to be sent to the system operators before serious problems occur. Figure 7.15 shows the information flow of system monitoring. In Figure 7.15, the SCADA system is not the only system to deliver the system monitoring function, smart metering and field maintenance crews are also integrated, for example, smart meters may be able to send 'last gasp' information when they lose their electricity supply.

System monitoring compares the measured data against their normal values or limits. Any abnormal change in the real-time measurements generates an event that triggers automatic control functions or notify the DNOs. In practice, deadbands are usually used to filter out normal and small fluctuations of the measured data, as shown in Figure 7.16.

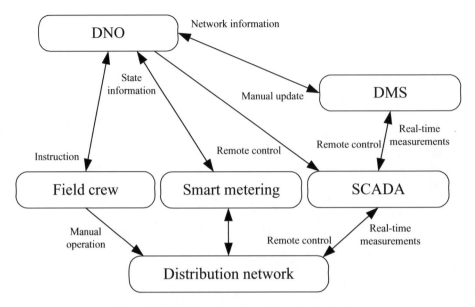

Figure 7.15   Information flow of system monitoring

7.4.2   *System operation*

7.4.2.1   Network reconfiguration

Distribution networks are normally constructed as a meshed network but are operated in a radial manner with normally open points. The network configuration can be varied through changing the open/close status of switchgear, manually or automatically. The main objectives of network reconfiguration are:

1. Supply restoration: This optimally restores electricity to customers using alternative sources. The Application is part of the fault location, isolation and supply restoration function.

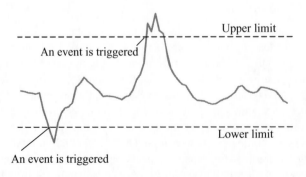

Figure 7.16   Deadbands used by the system monitoring functions

2. Active power loss minimisation at a given time or energy loss minimisation over a period of time.
3. Load balancing between different feeders or transformers and equalising voltages.

The methods used for network reconfiguration include those based on practical experience, mathematical methods, computational intelligence-based methods (for example, artificial neural networks, genetic algorithm, fuzzy logic), and hybrid methods which combine two or more methods.

7.4.2.2   Volt/VAR control

Volt/VAR control is used to improve voltage profiles and minimise network losses. This Application can be formulated as a multi-objective optimisation problem. Volt/VAR control calculates the optimal set points of the voltage controllers of OLTCs, voltage regulators, DER, power electronic devices, capacitor status and demand response. It optimises system operation by either following different objectives at different times or considering conflicting objectives together in a weighted manner.

7.4.2.3   Relay protection re-coordination

This Application adjusts relay protection settings in real time based on predetermined rules. This is accomplished through analysis of relay protection settings and operational modes of circuit breakers (that is, whether the circuit breaker is in a single-shot or recloser mode), while considering the real-time network connectivity, co-ordination with DER and MicroGrids, and weather conditions. This Application is called into use after feeder reconfiguration or when the bad weather is expected.

7.4.2.4   Operation of DER

The integration of DER operation to the DMS has a large impact on the performance of a smart distribution network. This integration depends on the interface between the DER and the distribution system. For example, a large DER unit may be integrated with the distribution system directly or through a power electronic interface or a large number of DER units may be connected through a MicroGrid [8] or a Cell.

Compared to a central generator, there are significant differences in the way that an individual DER is connected and controlled. The presence of DERs in the distribution network can significantly alter the flow of fault currents and change the source of ancillary services, so the operation of DER needs to be integrated to the DMS to guarantee reliable system operation.

A MicroGrid (Figure 7.17) can be defined as a low voltage electrical network of small modular distributed generating units (whose prime movers may be photovoltaics, fuel cells, micro turbines or small wind generators), energy storage devices and controllable loads. The integration of DMS and MicroGrids can be implemented through setting up the links between the DMS and the MicroGrid Central Controllers (MGCC).

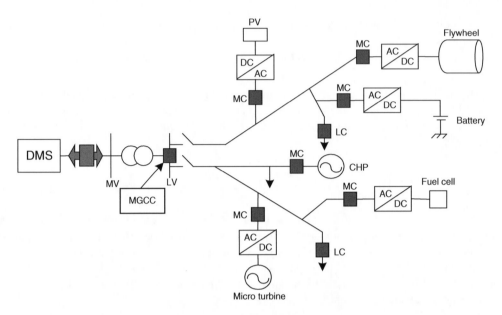

Figure 7.17   Integration of MicroGrids to DMS through MGCC

The cell concept considers a larger area of distribution network. There may be different cells at different levels of voltage which will have slightly different scope with regard to their technical requirements and what DER they have to control. Similar to MicroGrids, the integration of DMS and cells can be implemented through setting up links between the DMS and the cell controllers.

7.4.3   System management

The Automated Mapping (AM), Facilities Management (FM), and Geographic Information System (GIS) functions act as an integrated platform which links the automated digital maps of utility infrastructure to databases. There are two major components of an AM/FM/GIS system; the graphical component and the database component.

The graphical component deals with graphical data of various types of real world entities or objects represented graphically by shapes or geometries. The database component stores the attribute data for the real-world entities that need to be captured or managed as a part of the digitisation process. Relational database are usually used in practical applications.

7.4.4   Outage management system (OMS)

The OMS is a system which combines the trouble call centre and DMS tools to identify, diagnose and locate faults, then isolate the faults and restore supply. It provides feedback to customers that are affected. It also analyses the event and maintains historical records of the

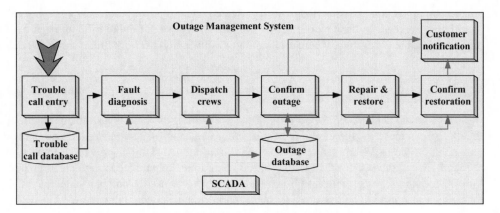

Figure 7.18   Information flow of the OMS

outage as well as calculating statistical indices of interruptions. The information flow of an OMS is shown in Figure 7.18.

Outage management is important in distribution networks with goals (and sometimes penalties) to restore the supply to a faulted section of the network within a period of time. The main functions of each part of OMS are as follows.

7.4.4.1   Fault identification

Fault identification is based on customer calls through telephone voice communication. It may also use automatic voice response systems (Computer Telephony Integration – CTI), automatic outage detection/reporting system, or SCADA detection of circuit breaker trip/lockout.

7.4.4.2   Fault diagnosis and fault location

Fault diagnosis and fault location are carried out based on the grouping of customer trouble calls using reverse tracing of the electrical network topology. It determines the protective device that is suspected to be open, for example, fuse, sectionaliser, recloser, or substation circuit breaker. Automatic feeder switching is also taken into account. The extent of the suspected outage will be calculated including the number of customers affected and the priority of the affected customers. Confirmation or modification of the fault diagnosis and its location is based on feedback from field crews.

Utilities with limited penetration of real-time control but good customer and network records use a trouble call approach, whereas those with good real-time systems are able to use direct measurements from automated devices. The trouble call solution is widely used in the United States for medium voltage networks. The lower voltage (secondary) feeder system is limited with, on the average, less than 6 and 10 customers being supplied from one distribution transformer. This system structure makes it easier to establish the customer-network link. In contrast, European systems with very extensive secondary systems (up to 400 consumers per distribution transformer) concentrate on implementing SCADA. Any MV fault would be cleared by protection and knowledge of the affected feeder known before any customer calls could be correlated. In this environment, to be truly effective, a trouble call approach has to operate from the LV system, where establishing the customer–network link is more difficult.

7.4.4.3 Supply restoration

Remedial action depends on the severity of the problem. If the fault is a simple problem, the field crew can make the repair and restore supplies in a short time. If the fault causes a major outage, after the isolation of the faulted area, the un-faulted portions will be restored using normally open points. The OMS tracks partial restorations. Automated fault detection, isolation, restoration schemes with feeder automation are widely used. Computer-aided modelling of crews is also used to help to analyse the capabilities, tools, equipment and workload.

7.4.4.4 Event analysis and recording

Any outage event will be analysed and the information kept as a historical record to record the cause, number of customers affected and duration. Such information is used for calculating performance statistics, for example, Customer Interruptions (CI) and Customer Minutes Lost (CML) as well as for planning/budgeting maintenance activities, for example, condition-based maintenance.

It is anticipated that smart metering will enhance the OMS function. The benefits from integrating smart metering and outage management are derived from crew and dispatcher efficiency savings, reduction in restoration costs and reduction of outage durations. Figure 7.19 shows the integration of smart metering and the OMS.

The last gasp messages from smart meters can be used as an input to the OMS. Fault diagnosis and fault location algorithms will operate more efficiently and effectively with such additional information. An OMS should consider a last gasp message in the same way as

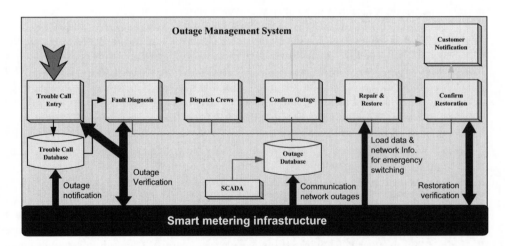

Figure 7.19 Integration of smart metering and the DMS

a customer phone call. Many OMS systems today require calls from less than 15 per cent of customers affected by an outage to predict the interruption device location accurately. Therefore, a 15– 20 per cent success rate of last gasp messages is thought to be adequate.

References

[1] Grainger, J.J. and Stevenson, W.D. (1994) *Elements of Power Systems Analysis*, McGraw-Hill, Maidenhead.
[2] Weedy, B. and Cory, B.J. (2004) *Electric Power Systems*, John Wiley and Sons, New York.
[3] Jenkins, N., Ekanayake, J.B. and Strbac, G. (2010) *Distributed Generation*, Institution of Engineering and Technology, Stevenage.
[4] Arrillaga, J. and Watson, N.R. (2001) *Computer Modeling of Electrical Power Systems*, John Wiley & Sons Ltd, New York.
[5] Cormen, T.H., Leiserson, C.E., Rivest, R.L. and Stein, C. (2001) *Introduction to Algorithms*, MIT Press and McGraw-Hill, New York.
[6] Kersting, W.H. (2001) *Distribution System Modelling and Analysis*, CRC Press, New York.
[7] Abur, A. and Exposito, A.G. (2004) *Power System State Estimation*, Marcel Dekker, Inc., New York.
[8] Hatziargyriou, N., Asano, H., Iravani, R. and Marnay, C. (2007) Microgrid: An overview of ongoing research, development and demonstration projects. *IEEE Power and Energy Magazine*, **5**(4), 78–94.

8

Transmission System Operation

8.1 Introduction

Transmission systems in many countries are facing ever more demanding operating conditions with increasing penetrations of renewable energy generation, larger flows and greater cross-border trading of electricity. The variability of the power output of renewable energy sources and unplanned flows through transmission grids are causing difficulties for the system operators, who are responsible for maintaining the stability of the system. Excessive power flows in transmission circuits and large variations in busbar voltages may arise during steady-state operation so that when faults and network outages occur, they can lead to system collapse.

In order to aid the transmission system operators to monitor, control, and optimise the performance of generation and transmission systems, a suite of Applications collected into an Energy Management System (EMS) is used. As the monitoring and control functions for EMS are often provided by SCADA, these systems are also referred to as EMS/SCADA. The EMS is normally located in the System Control Centre and effective real-time monitoring and remote control exist between the Control Centre and the generating stations and transmission substations.

With the growing availability of measurements from Phasor Measurement Units (PMUs), it is expected that, in future, PMU measurements will be integrated with EMS. However, at present, PMUs are mainly incorporated into separate Wide Area Applications. It is expected that EMS and Wide Area Applications will coexist separately for some time. Figure 8.1 shows how different data sources feed into Applications. Good visualisation tools are also important to represent information in an effective manner.

8.2 Data sources

8.2.1 IEDs and SCADA

As discussed in Chapter 6, IEDs receive measurements and status information from substation equipment and pass it into the process bus of the local SCADA. The substation SCADA

Smart Grid: Technology and Applications, First Edition.
Janaka Ekanayake, Kithsiri Liyanage, Jianzhong Wu, Akihiko Yokoyama and Nick Jenkins.
© 2012 John Wiley & Sons, Ltd. Published 2012 by John Wiley & Sons, Ltd.

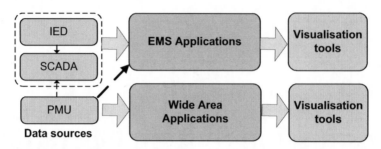

Figure 8.1   A structure of an EMS/SCADA and Wide Area Applications

systems are connected to the Control Centre where the SCADA master is located and the
information is passed to the EMS Applications.

8.2.2   *Phasor measurement units*

Figure 8.2 shows a PMU. It measures 50/60 Hz sinusoidal waveforms of voltages and currents at a high sampling rate, up to 1200 samples per second and with high accuracy. From
the voltage and current samples, the magnitudes and phase angles of the voltage and current signals are calculated in the phasor microprocessor of the PMU. As the PMUs use the
clock signal of the Global Positioning System (GPS) to provide synchronised phase angle
measurements at all their measurement points, the measured phasors are often referred to as
synchrophasors [1].

Figure 8.3 shows voltage synchrophasors at the two ends of an inductive transmission line.
The sinusoidal waveform of the voltage is expressed as:

$$v_i\,(t) = V_{m\_i}\sin\,(\omega t + \phi_i)$$

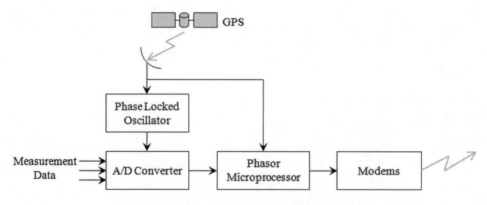

Figure 8.2   Phase Measurement Unit (PMU) device

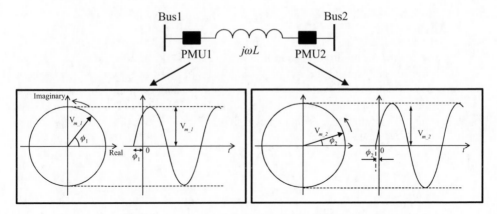

Figure 8.3   Waveforms and phasors of busbar voltages

where:

  i is the bus number at each end of the line (1 or 2)

  V_{m_i} is the peak value.

The voltage synchrophasor is given by:

$$\mathbf{V_i} = V_{m\_i}e^{j\phi_i} = \sqrt{2}V_{\text{rms}\_i}e^{j\phi_i}$$

where V_{rms_i} is the rms value of the voltage magnitude.[1]

  Synchrophasors measured at different parts of the network are transmitted to a Phasor Data Concentrator (PDC) at a rate of 30–60 samples per second. Each PDC sends the data that is collected to a super PDC where there is Application software for data visualisation, storing the data in a central database and for integration with EMS, SCADA and Wide Area Application systems (Figure 8.4).

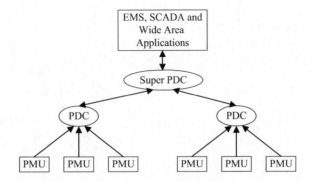

Figure 8.4   An example of a PMU connection

---

[1] Phasors are often expressed based on rms magnitude values.

Example 8.1

The power flow on the transmission line in Figure 8.5 is 5 pu and the voltage at both busbars is 1.0 pu. The system frequency is 50 Hz. The power flow is estimated using the phase difference between busbars 1 and 2, that is using $\phi_1 - \phi_2$. The measurement of the phase angle ϕ_1 has a time stamp error of 0.1 ms and that of the phase angle ϕ_2 is zero. Find the error in the estimated power flow.

Bus 1   Bus 2

PMU 1 $j0.1$ PMU 2
$V_1 = 1.0$ $V_2 = 1.0$

Figure 8.5   Figure for Example 8.1

Answer

The phase angle error $\Delta\phi$ is derived from the time stamp error of 0.1 ms and is given by:

$$\Delta\phi = \frac{0.1}{20} \times 2\pi \cong 0.0314 \, \text{rad}$$

The phase angle difference $\phi_1 - \phi_2$ is calculated in the following way:

$$P = \frac{V_1 V_2}{X} \sin(\phi_1 - \phi_2) = 10 \sin(\phi_1 - \phi_2) = 5 \, \text{pu}$$

$$\therefore \phi_1 - \phi_2 = \frac{\pi}{6}$$

The estimated power flow error ΔP is shown in Figure 8.6 and is given by:

$$\Delta P = \frac{V_1 V_2}{X} \sin(\phi_1 - \phi_2 + \Delta\phi) - \frac{V_1 V_2}{X} \sin(\phi_1 - \phi_2)$$

$$= 2 \times \frac{V_1 V_2}{X} \cos\left(\phi_1 - \phi_2 + \frac{\Delta\phi}{2}\right) \sin\left(\frac{\Delta\phi}{2}\right)$$

$$\approx 2 \times \frac{V_1 V_2}{X} \cos(\phi_1 - \phi_2) \left(\frac{\Delta\phi}{2}\right)$$

$$= 2 \times \frac{1}{0.1} \times \cos\left(\frac{\pi}{6}\right) \times \frac{0.0314}{2} = 0.272 \, \text{pu}$$

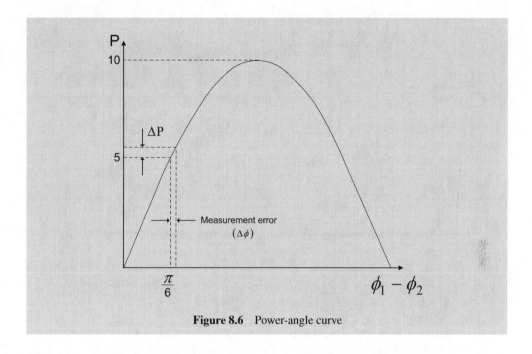

Figure 8.6 Power-angle curve

8.3 Energy management systems

Energy Management Systems (EMS) were designed originally at a time when the electrical power industry was vertically integrated and had centralised communications and computing systems. With deregulation of the power industry and the development of the Smart Grid, decision-making is becoming decentralised, and coordination between different actors in various markets becomes important.

A number of international standards are emerging (or are in place) for the abstract description of systems and services. The abstract description of interfaces is facilitated through the use of the Unified Modelling Language (UML) which provides an object-orientated representation of the power system. Use of standardised data models allows Applications from different vendors to be integrated, thus reducing the necessity for data wrappers. Today, IEC 61970 is the commonly used standard for EMS systems.

A typical EMS system configuration is shown in Figure 8.7. System status and measurement information are collected by the Remote Terminal Units (RTUs) and sent to the Control Centre through the communication infrastructure. The front-end server in the EMS is responsible for communicating with the RTUs and IEDs. Different EMS Applications reside in different servers and are linked together by the Local Area Network (LAN).

EMS Applications include Unit Commitment, Automatic Generation Control (AGC), and security assessment and control. However, an EMS also includes Applications similar to those of a DMS and most of the tools used in a DMS such as: topological analysis, load forecasting, power flow analysis, and state estimation (Chapter 7).

The purpose of Unit Commitment within a traditional power system is to decide how many and which generators should be used and to allocate the sequence of starting and shutting

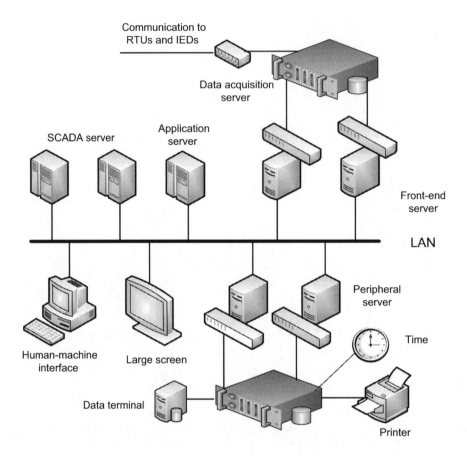

Figure 8.7 A typical EMS system configuration

down generators. The inputs to the Unit Commitment Application include past load demand records, generator cost and emission functions, and accurate load forecasts. However, in a number of countries, market-based trading of electrical energy has superseded classical Unit Commitment. Then a number of forward markets operate to determine which generators should run. The system operator only arranges generator dispatch and the balancing of supply and demand during a final period (1 hour before real time in Great Britain).

Similarly, in a power system, AGC carries out load frequency control and economic dispatch. Load frequency control has to achieve three primary objectives to maintain: (1) system frequency; (2) power interchanges with neighbouring control areas; and (3) power allocation between generators at the economic optimum. AGC also performs functions such as reserve management (maintaining enough reserve in the system) and monitoring/recording of system performance. In a deregulated power system, many of these functions are managed through markets rather than directly by the system operator through an AGC.

Security assessment and control may be understood using the widely used framework proposed by Dy Liacco [2] (Figure 8.8). This Application exercises control to keep the power

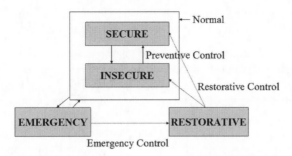

Figure 8.8 The Dy Liacco Framework for security assessment and control

system in a secure state. The Dy Liacco framework considers the power system as being operated under two types of constraint: load constraints (load demand must be met), and operating constraints (maximum and minimum operating limits together with stability limits should be respected). In the normal state, both these constraints are satisfied.

The security assessment and control Application includes; security monitoring, security analysis, preventive control, emergency control, fault diagnosis and restorative control. The tools required include:

- network topology analysis
- external system equivalent modelling
- state estimation
- on-line power flow
- security monitoring (on-line identification of the actual operating condition – secure or insecure)
- contingency analysis.

When the system is insecure, security analysis informs the operator which contingency is causing insecurity and the nature and severity of the anticipated emergency.

Besides these EMS functions, a training tool, the Dispatcher Training Simulator, is embedded within an EMS. Dispatch Training Simulators were originally created as a generic Application to introduce operators to the electrical and dynamic behaviour of a power system. Today, they model the actual power system being controlled with reasonable fidelity and are integrated within the EMS to provide a realistic environment for operators and dispatchers to practise normal, everyday operating tasks and procedures as well as experiencing emergency operating situations. Various training activities can be practised safely and conveniently with the simulator responding in a manner similar to the actual power system.

8.4 Wide area applications

Wide-Area Measurement Systems (WAMS) are being installed on many transmission systems to supplement traditional SCADA. They measure the magnitudes and phase angle of busbar voltages as well as current flows through transmission circuits. This information, measured over a wide area, is transmitted to the Control Centre and is used for:

1. *Power system state estimation*: Since the phasor data is synchronised, the magnitudes and phase angles of voltages at all busbars in the grid can be estimated using a state estimation algorithm. These estimates can then be used to predict possible voltage and angle instabilities as well as to estimate system damping and vulnerability to small-signal oscillation.
2. *Power system monitoring and warning*: The phasor data allows the operating conditions of the power system to be monitored on a real-time basis, system stability to be assessed and warnings generated.
3. *Power system event analysis*: Synchronised phasor data of high accuracy is available before and after a fault or other network incident. This enables the system operators to study the causes and effects of faults and take countermeasures against subsequent events.

In future, it is anticipated that these Applications will be integrated into a Wide Area Monitoring, Protection and Control (WAMPAC) system [3]. Some examples of possible future WAMPAC schemes include:

1. To initiate actions to correct the system once a voltage, angle or oscillatory instability has been predicted. This may include switching of generators and controlling devices such as the Flexible AC transmission Systems (FACTS), the Power System Stabilisers (PSS) and HVDC converters.
2. To generate emergency control signals to avoid a large-scale blackout (for example, through selective shedding of load or temporary splitting of the network) in the event of a severe fault.

A configuration of the WAMPAC is shown in Figure 8.9. The PMU (or synchrophasor) measurements collected from the different part of the network and state estimation are used for

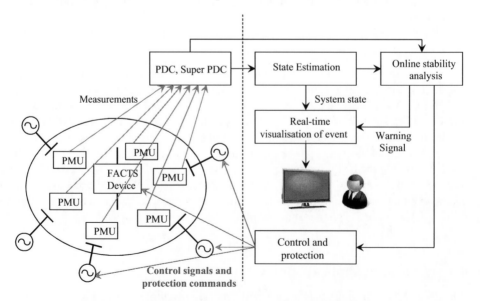

Figure 8.9   Simplified representation of WAMPAC

online stability analysis. When an event occurs, its location, time, magnitude (total capacity of generator or transmission lines outage) and type (generator outage or transmission line outage) are first identified. Real-time visualisation of the event allows it to be replayed several seconds after it occurs. The future system condition is then analysed using the information that has been gathered. An on-line stability assessment algorithm continuously assesses the system to check whether the system is still stable and how quickly the system would collapse if it became unstable. If instability is predicted, then the necessary corrective actions to correct the problem or to avert system collapse are taken.

8.4.1 On-line transient stability controller

In [4], an on-line Transient Stability Controller was discussed that would trip a number of generator units when a fault occurs on extra high voltage transmission lines (500 kV and 275 kV) in order to prevent transient instability. The operation of the on-line Transient Stability Controllers is described in Figure 8.10. Using PMU data and results from the state estimator, transient stability analysis is carried out repeatedly (typically every 5 minutes) and the generator units to be shed if a fault occurs are determined. After a fault occurs, the fault is compared with those of the contingencies identified pre-fault and it determines the generator units to be shed. Then a signal is sent to the local control units to shed the identified generator units.

8.4.2 Pole-slipping preventive controller

When a severe fault occurs in a power system, this controller would predict unstable conditions of the power system and rapidly trips an appropriate number of generator units or splits the system into two subsystems in order to prevent pole slipping.

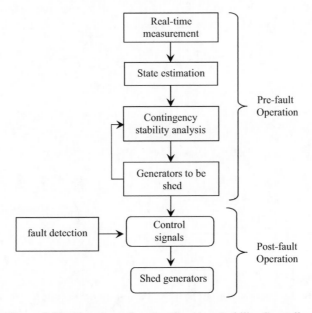

Figure 8.10   Functions of on-line Transient Stability Controller

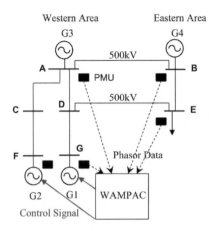

Figure 8.11   Pole-slipping prevention controller for 500 kV Transmission Network

To illustrate the operation of this controller, two areas, the Western and the Eastern systems, interconnected through a 500 kV transmission lines are shown in Figure 8.11. Due to a fault, line DE was tripped and it was found that the phase angle between the Western and Eastern areas had increased. It was predicted by the WAMPAC that the large phase angle swing might cause synchronism between the two areas to be lost. Therefore, a control signal was sent to trip generator G1 (or G2).

Example 8.2

Obtain the phase angle difference between voltage $\mathbf{V_{G1}}$ and $\mathbf{V_{G2}}$ using the data shown in Figures 8.2 and 8.3. In Figure 8.4, $\mathbf{V}_{G1}$ and $\mathbf{V}_{G2}$ are the sampled data of $\mathbf{V_{G1}}$ and $\mathbf{V_{G2}}$ at time t_0 and $\mathbf{V}'_{G1}$ and $\mathbf{V}'_{G2}$ are the sampled data of $\mathbf{V_{G1}}$ and $\mathbf{V_{G2}}$ at time $(t_0 - \pi/2\omega)$.

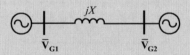

Figure 8.12   Figure for Example 8.2

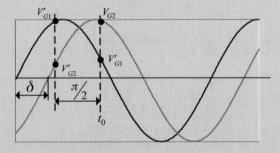

Figure 8.13   Instantaneous values of bus voltage waveforms for Example 8.2

Answer

At $t = t_0$:

$$V_{G1} = V_{m1} \sin \omega t_0 \text{ and}$$
$$V_{G2} = V_{m2} \sin (\omega t_0 - \delta)$$

where V_{m1} and V_{m2} are peak values.

At $t = \left(t_0 - \frac{\pi}{2\omega}\right)$:

$$V'_{G1} = V_{m1} \sin (\omega t_0 - \pi/2) = V_{m1} \cos \omega t_0 \text{ and}$$
$$V'_{G2} = V_{m2} \sin (\omega t_0 - \pi/2 - \delta) = V_{m2} \cos (\omega t_0 - \delta)$$

From trigonometry:

$$\sin \delta = \sin \omega t_0 \cos (\omega t_0 - \pi/2) - \cos \omega t_0 \sin (\omega t_0 - \pi/2) = V_{G1} V'_{G2} - V_{G2} V'_{G1}$$
$$\cos \delta = \sin \omega t_0 \sin (\omega t_0 - \pi/2) + \cos \omega t_0 \cos (\omega t_0 - \pi/2) = V_{G1} V_{G2} + V'_{G2} V'_{G1}$$

Phase angle difference is given by:

$$\delta = \tan^{-1} \left[\frac{V_{G1} V'_{G2} - V_{G2} V'_{G1}}{V_{G1} V_{G2} + V'_{G1} V'_{G2}} \right]$$

8.5 Visualisation techniques

In contrast to distribution networks, a transmission system has a large number of sensors and will have good communication infrastructure that is able to transmit real-time data. However, decision-making at transmission level is complex and this complexity will increase as:

- the use of renewable energy increases;
- cross-border flows become larger;
- new data becomes available, for example, by PMUs.

Visualisation is now one of the core elements in an EMS and essential to manage large volumes of data. It acts as the interface between the EMS Applications and the system operators. Visualisation can speed up the decision-making of the system operators and improve the quality of their choice of action.

Visualisation techniques convert data and information into geometric or graphic representations. However, more than simple presentation, effective visualisation enables the display of the calculation processes and results in an effective way. It provides an intuitive way to express the large quantity of complex data so that control engineers can respond to system emergencies and contingencies in the short time that is available. Techniques used for visualisation include [5]:

1. *Two-dimensional (2-D) presentation, 3-dimensional (3-D) presentation, virtual reality and animation*: These highlight key information in a manner easily recognised by the operator.

2. *Multi-resolution modelling*: This provides a variety of scales of resolution that helps to prevent information overload while allowing users to find highly detailed information when needed. It allows the users to view varying levels of detail, and zoom into areas of specific interest, such as line outages. It supports integration with other sources of information, such as live cameras or thermometers.
3. *Faster than real-time performance*: This visualisation tool utilises increased computational power to analyse systems faster than real-time speeds, providing forecasts, contingency analyses, and suggested courses of action.
4. *Geographically integrating wide areas*: A GIS platform is used for the wide area visualisation.

8.5.1   Visual 2-D presentation

Extensive information is available in the literature regarding the design and use of 2-D presentations to visualise data and information [5]. These insights have been used to upgrade many power system control rooms. In 2-D presentations, images can be viewed and explored interactively on a conventional computer monitor display. Keyboards or a mouse can be used to rotate images in all directions, move to different positions, zoom, assemble or disassemble the components making up an object, and demonstrate operation by use of animation. Labels can be used to provide information of the objectives and their components. Hyperlinks can be used to link to other materials, for example, videos and internet websites.

Figure 8.14 (also refer to Plate 3) gives an example of a visual 2-D presentation. This shows the voltage of all nodes in a network. All nodes are ranked by their voltage from maximum to minimum and shown sequentially using a black dot. The busbar name appears next to each black dot. The concentric circles bounded by different colours represent different voltages. The concentric circle in the middle (dashed) shows the nominal voltage and the voltage excursion is marked at the boundary of coloured circles. A blue line connecting each

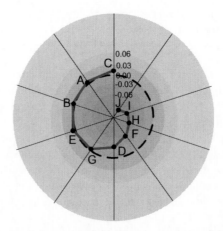

Figure 8.14   Visual 2-D presentations of nodal voltages. *Source:* Courtesy of Tianda Qiushi Power New Technology Co., Ltd, China

node voltage shows that the nodal voltage is normal and the red colour line shows that the nodal voltage is too low.

8.5.2   Visual 3-D presentation

There are two main methods for representing digital objects in three dimensions: surface rendering and direct volume rendering. Surface rendering is the method most commonly used in movies and video games. This method uses an indirect geometry-based technique to create a series of polygons (typically triangles) meshed together end-to-end to form the surface of an object. The polygons are then filled with the necessary colours to represent the effects of lighting and object properties. The results can have a photo-realistic appearance. In this approach the model is hollow in its core [5].

In contrast, the volume graphics method is based on capturing a series of discrete points in the space describing the object of interest. These points, or volume elements, are pixels with X, Y and Z coordinates. Therefore, models created in this way have both a surface and a core. Users can assign attributes such as various system operational limits to each element and then render by attributes. Users can take apart models, manipulate individual parts, and modify information all in real time. As with surface rendering, the resulting objects can have a photorealistic appearance.

Figure 8.15 (also refer to Plate 4) gives an example of visual 3-D presentation, which visualises the operating condition and limits of two interconnected control areas using the volume graphics method. There are several interconnected circuits between the two control areas, any three of them can be chosen by the user to be the three coordinates in Figure 8.15. Labels indicate power flow in each interconnection in MW. The white point represents the current operating point and the white surface represents the operating limit of three interconnectors. Through such 3-D visualisation, the distance between the current operating point and the system limits can easily be seen by the system operator.

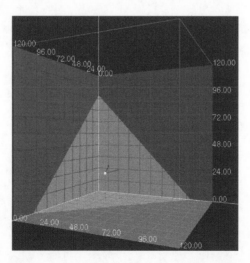

Figure 8.15   Visual 3-D presentations of a cut-set voltage stability region. *Source:* Courtesy of Tianda Qiushi Power New Technology Co., Ltd, China

References

[1] *IEEE C37.118: Standard for Synchrophasors for Power Systems*, IEEE, 2005.

[2] Dy Liacco, T.E. (1967) The adaptive reliability control system, *IEEE Transactions on Power Apparatus and Systems*, **PAS**-86 (5), 517–531.

[3] *Network Protection and Automation Guide: Protective Relays, Measurements and Control*, May 2011, Alstom Grid, Available from http://www.alstom.com/grid/NPAG/ on request.

[4] Ota, H., Kitayama, Y., Ito, H. *et al.* (1996) Development of transient stability control system (TSC system) based on on-line stability calculation. *IEEE Transactions on Power Systems*, **11**(3), 1463–1472.

[5] Wright, H. (2007) *Introduction to Scientific Visualization*, Springer, New York.

Part III

Power Electronics and Energy Storage

9

Power Electronic Converters

9.1 Introduction

The future power system increasingly will include more controllable power electronic devices to make the best use of existing circuits, maintain flexibility and optimum operation of the power system, and to facilitate the connection of renewable energy resources at all voltage levels. Figure 9.1 illustrates a future power system that is rich in power electronics.

Current Source Converter High Voltage DC (CSC-HVDC[1]) is presently used for the connection of asynchronous power systems (for example, 50–60 Hz), for long overhead line transmission and submarine cable circuits as well as for the connection of geographically extensive or weak systems [1–3]. It is anticipated that CSC-HVDC connections will be increasingly used in future for inter-country and inter-state connections.

Voltage Source Converter HVDC (VSC-HVDC[1]) is used for offshore wind farm connections. Offshore wind farms up to 50–80 km from shore can be connected to the terrestrial grid using an AC connection. However, the submarine cables generate significant reactive power limiting the distance over which an AC cable connection can be used. The decision on whether to use AC or DC depends on the cable route length, the number of cables required to transmit the wind farm output, acceptable losses and capital costs.

Flexible AC Transmission Systems (FACTS) are used to increase the power transfer capability of existing AC lines, to control steady state and dynamic power flow through an AC circuit, to control reactive power and voltage and to enhance voltage and angle stability [4, 5]. Shunt-connected power electronic devices using Thyristors, for example, Static Var Compensators (SVCs) using Thyristor Controlled Reactors or Thyrsistor Switched Capacitors are already widely used. Other FACTS devices, given in Table 9.1, have been demonstrated. Other than some commercial applications of STATCOMs, TCSCs, and TSSCs, the use of other FACTS devices (IPFC and UPFC) has yet to become widespread.

The Static Var Compensator (SVC) and STATCOM provide a rapid injection of reactive power. They are used with wind farms for reactive power support to satisfy Grid Connection codes (which define the requirements for the connection of generation and loads to an electrical

---

[1] The Current Source Converter HVDC is described in Section 11.3.1 and Voltage Source Converter HVDC is described in Section 11.3.2.

Smart Grid: Technology and Applications, First Edition.
Janaka Ekanayake, Kithsiri Liyanage, Jianzhong Wu, Akihiko Yokoyama and Nick Jenkins.
© 2012 John Wiley & Sons, Ltd. Published 2012 by John Wiley & Sons, Ltd.

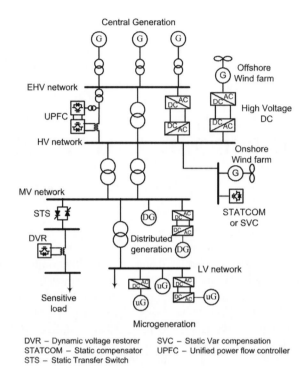

Figure 9.1   Future power system

Table 9.1   Different FACTS devices and their applications

Shunt or series	FACTS device	Switches used	Applications Damping oscillations, transient and dynamic stability, voltage stability are common applications. Other unique applications are listed below
Shunt device	Static Var Compensator (SVC)	Thyristor	Voltage control, Var compensation
	Static Compensator (STATCOM)	IGBT, IGCT	
Series device	Thyristor Controlled Series Compensator (TCSC)	Thyristor	Power flow control
	Thyristor Switched Series Compensator (TSSC)	Thyristor	
Shunt and Series device	Integral Power Flow Controller (IPFC)	IGBT, IGCT	Active and reactive power control
	Unified Power Flow Controller (UPFC)	IGBT, IGCT	

network). The TCSC and TSSC are used with long transmission lines to increase the power transfer capability, to control the power flows and to damp oscillations.

Power electronic interfaces are used with distributed generators and micro-generators for variable speed operation, to control their active and reactive power output and terminal voltage. In photovoltaic (PV) and energy storage applications, the power electronic interfaces convert DC into AC suitable for grid connection [6]. In wind and hydro turbine applications, the power electronic interfaces are used for variable speed operation, thus decoupling the turbine speed from the frequency of the grid [7, 8]. This enables the maximum power to be extracted from the turbines and also reduces their mechanical loadings.

The proliferation of non-linear and sensitive loads increases the need for controllable devices to maintain power quality. A range of power electronic devices are now in operation to reduce the frequency and duration of power interruptions, to maintain the network voltage within limits and to minimise harmonic distortion [9]. Some of these devices are Dynamic Voltage Restorers (DVR), Active Filters, and Uninterruptable Power Supplies (UPS).

This chapter describes Current Source Converters (CSC) and Voltage Source Converters (VSC) that are used in many of these devices. In Chapter 10 ('Power Electronics in the Smart Grid'), controllable power electronic interfaces in energy supply and demand are discussed. Technologies for circuit control such as HVDC and FACTS are addressed in Chapter 11 ('Power Electronics for Bulk Power Flows').

9.2   Current source converters

The rapid development of semiconductor devices and associated control techniques has allowed a number of applications of switching power converters in the electric power system. Two types of power converter, the Current Source Converter (CSC) and the Voltage Source Converter (VSC) are in use [10]. In a CSC, the DC side current is kept constant with a small ripple using a large inductor, thus forming a current source on the DC side. The direction of power flow through a CSC is determined by the polarity of the DC voltage while the direction of current flow remains the same. At present, the CSC is used mainly in high power applications, particularly for HVDC transmission with a capacity of up to 7000 MW for a single link. In a CSC, the power electronic switches (thyristor valves) are turned on by control circuits but switch off through natural commutation when the current through them drops to zero.

Figure 9.2 shows the circuit configuration of a CSC. Six thyristors are used to form this circuit and the midpoint of each thyristor limb is connected to a three-phase supply. During the period when the a-phase voltage, v_a, is most positive, thyristor T_1 can be turned on and during the period when v_b is most positive, thyristor T_3 can be turned on. This pattern is repeated for the other phase. The lower thyristor in the a-phase, that is, T_4, can be turned on during the time the a-phase voltage is most negative. The DC voltage, V_d, depends on the upper and lower thyristors that are conducting. For example, when thyristors T_1 and T_6 are conducting, V_d (the voltage of the positive DC conductor with respect to the negative) is the line-to-line voltage between the a and b phases, that is, V_{ab}.

Figure 9.3 (also refer to Plate 5) shows the DC voltage and thyristors that are conducting if they are turned on without a firing angle delay. The average value of the DC voltage can be obtained by dividing the shaded area by $\pi/3$ (the angle of the shaded area). Since V_d is

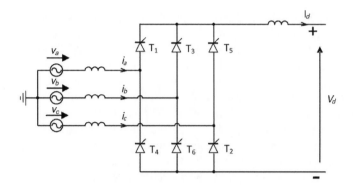

Figure 9.2 Current Source Converter

formed by the line-to-line voltage, its peak value is $\sqrt{2}V_{\text{LL}}$. So the average DC voltage is given by:

$$V_d = \frac{3}{\pi} \int\limits_{-\pi/6}^{\pi/6} \sqrt{2}V_{\text{LL}} \cos\theta \, d\theta = \frac{3\sqrt{2}}{\pi} V_{\text{LL}} \left[\sin\theta\right]_{-\pi/6}^{\pi/6}$$

$$= \frac{3\sqrt{2}}{\pi} V_{\text{LL}} = 1.35 V_{\text{LL}}$$

(9.1)

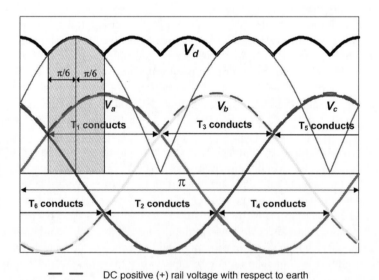

Figure 9.3 Operation of the CSC without firing angle delay

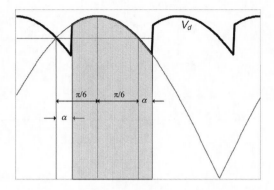

Figure 9.4   Operation of the CSC with firing angle delay

If thyristor T_1 is not triggered at the point when V_a becomes more positive than V_c or, in other words, if there is a firing angle delay in T_1, the previous thyristor on the upper side, that is T_5, continues to conduct until T_1 is turned on. Figure 9.4 shows the DC voltage with a firing angle delay of α. Now the shaded area is the integration of same cosine function, but from $-(\pi/6 - \alpha)$ to $(\pi/6 + \alpha)$. Therefore, the average DC voltage is given by

$$V_d = \frac{3}{\pi} \int\limits_{-(\pi/6-\alpha)}^{(\pi/6+\alpha)} \sqrt{2}V_{LL} \cos\theta \, d\theta$$

$$= \frac{3\sqrt{2}}{\pi} V_{LL} \left[\sin\theta \right]_{-(\pi/6-\alpha)}^{(\pi/6+\alpha)}$$

$$= \frac{3\sqrt{2}}{\pi} V_{LL} \cos\alpha = 1.35 V_{LL} \cos\alpha \qquad (9.2)$$

Since α can vary from $0°$ to $180°$, V_d can be varied from $+1.35V_{LL}$ to $-1.35V_{LL}$. When V_d is positive, power flows from the AC side to the DC side, thus the CSC acts as a rectifier. On the other hand, when V_d is negative, power flows from DC side to AC side, thus the CSC acts as an inverter.

If the inductor in the DC side is large, it can be assumed that the DC current, I_d is constant. Then whenever thyristor T_1 conducts, the a-phase AC side current is equal to I_d and whenever T_4 conducts, the a-phase AC side current is equal to $-I_d$ (see Figure 9.2). The a-phase AC side current and voltage are shown in Figure 9.5. A similar pattern is followed by the b and c phases. From Fourier analysis of the current, it may be shown that:

$$i_a = \frac{2\sqrt{3}}{\pi} I_d \left[\cos\omega t - \frac{1}{5}\cos 5\omega t + \frac{1}{7}\cos 7\omega t - \frac{1}{11}\cos 11\omega t + \frac{1}{13}\cos 13\omega t - \ldots.. \right]$$

$$(9.3)$$

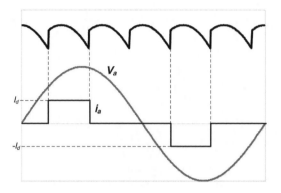

Figure 9.5   AC side current and voltage

From Equation (9.3):

$$\text{The peak value of the fundamental} = \frac{2\sqrt{3}}{\pi}I_d \tag{9.4}$$

$$\text{Therefore the rms value} = \frac{2\sqrt{3}}{\pi}\frac{1}{\sqrt{2}}I_d = \frac{\sqrt{6}}{\pi}I_d \tag{9.5}$$

Ignoring losses and equating power on the AC and DC sides: $3V_{\text{RMS}}I_{\text{RMS}}\cos\phi = V_dI_d$. Substituting for $V_{\text{RMS}} = V_{\text{LL}}/\sqrt{3}$ and I_{RMS} and V_d from Equations (9.5) and (9.2), the following equations can be obtained:

$$3 \times \frac{V_{\text{LL}}}{\sqrt{3}} \times \frac{\sqrt{6}}{\pi}I_d \cos\phi = \frac{3\sqrt{2}}{\pi}V_{\text{LL}}\cos\alpha \times I_d$$

$$\cos\phi = \cos\alpha$$

$$\therefore \ \phi = \alpha \tag{9.6}$$

As the angle α changes, the phase displacement between the AC side voltage and current of the CSC also changes. From Equation (9.6), the power factor angle is equal to α.

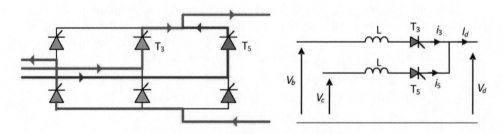

Figure 9.6   The commutation process

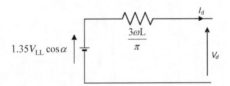

Figure 9.7   Equivalent circuit in rectifier mode

In the preceding section it is assumed that the commutation process is instantaneous. In other words, one thyristor stops conducting suddenly and transfers the current to the next thyristor immediately. However, this is not true as the system inductance does not allow the current through a thyristor to extinguish suddenly. Figure 9.6 shows commutation from T_3 to T_5 (that is, T_3 will stop conducting and T_5 will start conducting at this instance). As shown in Figure 9.6, V_d is no longer V_{ba} as there will be a circulating current between T_3 and T_5. This circulating current reduces the average value of the DC side voltage. It can be shown that $V_d = 1.35V_{LL}\cos\alpha - (3\omega L/\pi)I_d$ [1–3]. From this equation, a DC equivalent circuit of the rectifier, which is a DC source of magnitude $1.35V_{LL}\cos\alpha$ and a series resistance (not a physical resistor but an equivalent resistance to represent the commutation process) of $(3\omega L)/\pi$ (as shown in Figure 9.7), can be derived.

9.3   Voltage source converters

A converter where the DC side voltage is maintained as constant using a large capacitor is called a VSC [10, 11]. The VSC is widely used for low and medium power applications such as grid connection of micro-generation, renewable energy sources, and energy storage. However, the VSC is also used at power levels up to 1000 MW for HVDC transmission. The semiconductor devices that are used in VSC are rapidly increasing in size and reducing in cost. Therefore, it is anticipated that VSC technology will dominate high power DC applications in future. The VSC offers advantages such as freedom to operate with any combination of active and reactive power, the ability to operate in a weak grid and even black-start, fast acting control, the possibility of using voltage polarised cables and generating good sinusoidal wave-shapes. Hence it is anticipated that it will be the choice of future converters.

A VSC employs controllable switches where current can be controlled in the forward direction and an anti-parallel diode is provided for current flow in the reverse direction. The current in the forward direction can be switched on and off. Commonly used semiconductor switches include Metal Oxide Semiconductor Field Effect Transistor (MOSFET), Insulated Gate Bipolar Transistor (IGBT), Gate Turn-off Thyristor (GTO), and Insulated Gate Commutated Thyristor (IGCT). The current and voltage ratings of MOSFETs are limited and therefore they are only used for low power applications. IGBTs have been used in both low and medium power applications. As their current and voltage ratings increase, it is anticipated that IGBT switches will be used in high power applications (already a few HVDC projects of up to 500 MW are employing them). The IGCT is evolving as a device having lower on-state losses (compared to an IGBT) and faster switching times (compared to a GTO) and is used in some high power applications.

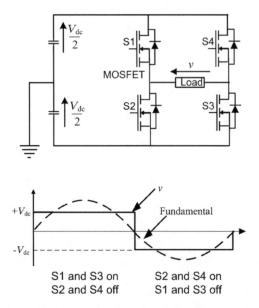

Figure 9.8   H-Bridge VSC with a square wave output

9.3.1   *VSCs for low and medium power applications*

In this book, low/medium power refers to applications in which the rating of the VSC is in the range of a few kWs to several MWs. Applications include inverters for PV and energy storage systems, back-to-back VSCs for wind power generators, active filters and DVRs. Some of these Applications use a single-phase VSC whereas others use a three-phase VSC.

9.3.1.1   **Single-phase voltage source converter**

The commonly used single-phase VSC is the H-bridge converter shown in Figure 9.8. The operation with square wave output is also shown in Figure 9.8. In this configuration the switch pairs (S1 and S3) and (S2 and S4) are turned on and off in a complementary manner. However, in order to avoid shoot-through of one leg (thus causing a high short circuit current), a small dead time is introduced between the turn-on signals of two sets of complementary switches. When the two switches S1 and S3 are on, the voltage across the load is V_{dc}, whereas when the two switches S2 and S4 are on, the voltage across the load is $-V_{dc}$. Even though the output is a square wave, its fundamental is sinusoidal as shown in Figure 9.8. The four switches could be MOSFETs (as shown in Figure 9.8) or IGBTs.

Due to the harmonics produced by the square wave voltage output, this simple switching strategy is only used with off-grid low power generators. In many applications a sine-triangular Pulse Width Modulation (PWM) technique is used to control the turn-on and turn-off times of the four switches. The switching instances are determined by comparing a sinusoidal modulating signal with a triangular carrier signal (see Figure 9.9). When the magnitude of the carrier is higher than the modulating signal, S2 and S4 are turned on. On the other hand, when the magnitude of the carrier is lower than the modulating signal, S1 and S3 are turned on.

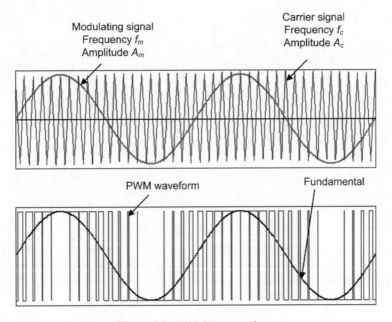

Figure 9.9   PWM output voltage

The PWM output voltage has a fundamental component and a series of harmonics. Harmonic frequencies depend on the frequency modulation index, m_f, which is defined as the ratio between f_c (frequency of the Carrier signal) and f_m (frequency of the Modulating signal). The harmonics are at frequencies $(pm_f \pm q)f_m$, where $p = 1, 2, 3 \ldots$. [10]. When p is odd then q is zero or even, that is if $p = 3$, harmonics will occur at $3m_f f_m$, $(3m_f \pm 2)f_m$, $(3m_f \pm 4)f_m$ and so on. When p is even then q is odd, that is if $p = 2$, harmonics will occur at $(2m_f \pm 1)f_m$, $(2m_f \pm 3)f_m$ and so on.

As each switch is turned on and off many times in a cycle, a VSC switched by a PWM technique produces more switching losses than a VSC that produces square waves. However, square wave VSC operation is less attractive due to the low order harmonics it generates and the filtering needed to control these harmonics.

Figure 9.10 shows the PWM pattern shown in Figure 9.9 for two different amplitude modulation indexes. As can be seen from Figure 9.10, the average of the PWM pattern in each time period T (period of triangular waveform) approximates to the fundamental. The peak value of the fundamental is approximately given by $(T_p/T)(V_d/2)$ (average of the PWM pattern time in the period closest to the peak). As can be seen from Figures 9.10a and 9.10b, time period T_p reduces with the modulation index. When $m_a = 1$, $T_p = T$ and the peak value of fundamental is equal to $(V_d/2)$. On the other hand, when $m_a = 0$, $T_p = 0$ and the peak value of fundamental is equal to zero. As described in [10], the peak value of the fundamental is approximately given by $(m_a V_d)/2$.

9.3.1.2   Two-level three-phase voltage source converter

The two-level three-phase VSC, also called a six-pulse VSC, shown in Figure 9.11 is essentially a three-limb configuration of two complementary switches. A number of different modulating

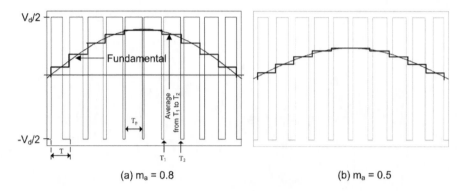

(a) $m_a = 0.8$ (b) $m_a = 0.5$

Figure 9.10   PWM output for two different modulation indexes

techniques can be used to generate the AC output [12, 13] but the sine-triangular PMW technique is a commonly used method. In sine-triangular PWM, the output of each phase is obtained by comparing the triangular carrier signal with three sinusoidal modulating signals which are out of phase by 120°.

9.3.1.3   Three-Level Three-Phase Diode Clamp Converter

Figure 9.12 shows a three-level three-phase diode clamp converter configuration. This is the simplest multilevel configuration; more levels could be added by increasing the number of series-connected switches. In this configuration, auxiliary devices are used to clamp the output terminal to the potential of the DC-link mid-point (O). This configuration is also called a Neutral-Point-Clamped (NPC) converter.

The lower trace of Figure 9.12 shows the a-phase output voltage with respect to the mid-point, O. The upper switches (S_{a1}, S_{a2}) are used to produce positive DC voltage at the output. The lower switches (S_{a3}, S_{a4}) are used to produce a negative DC voltage at the output. The middle switches (S_{a2}, S_{a3}) are used to clamp the output voltage to the potential of point O. Therefore, this switching arrangement can produce three output levels $+V_{dc}/2$, 0, $-V_{dc}/2$. A detailed discussion on switching states of the three-level converter can be found in [14].

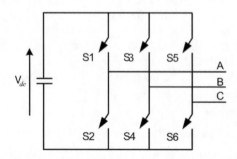

Figure 9.11   Three-phase VSC

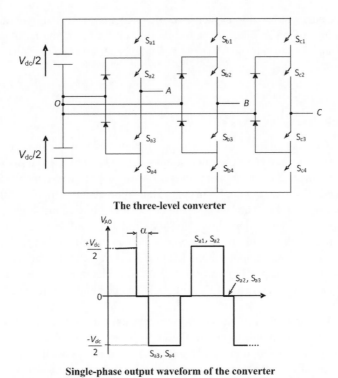

The three-level converter

Single-phase output waveform of the converter

Figure 9.12   Three-level three-phase diode clamp converter

During operation of the three-level VSC, the DC capacitor voltages should remain balanced. Different control strategies are employed for capacitor balancing [15].

9.3.2   *VSC for medium and high power applications*

9.3.2.1   Two-level three-phase voltage source converter

The two-level circuit shown in Figure 9.11 can be used for high power applications. In an application such as the HVDC, the voltages required (approx ±150 kV) are far above the rating of a single semiconductor device (approx 5 kV). Therefore, a large number of switches are connected in series to form a composite switch called a 'valve'. Within a valve all devices are switched together and therefore the principle of operation is the same as that described in Section 9.2.

One of the main problems of using a two-level VSC with PWM switching for high power applications is the high switching loss as each switch in the converter valves is turned on and off a large number of times per cycle of the 50/60 Hz. Also, with a series-connected string of switches it is difficult to ensure reasonable voltage sharing in the transient and steady state, so that the devices are not over-stressed. Unbalanced voltage sharing arises due to device parameter spread and gate drive delays. Passive snubbers may be used for voltage balancing.

A resistor–capacitor or resistor–capacitor–diode circuit is connected in parallel with each of the switches in a valve. Another method used for voltage balancing is to use an active gate control technique where gate charge is controlled to balance the voltage sharing of a series of switches [16].

9.3.2.2 Multi-level converters

Multi-level converters are an attractive choice for high power applications due to the reduced frequency of switching (this reduces switching losses). The output of the multi-level converter is a step-like output and by using many levels, a wave shape that closely resembles a sinusoid can be obtained. The main topologies in use are the diode-clamped topology and the capacitor-clamped topology [17, 18]. The Diode-Clamped Converter (DCC) topology is a generalisation of the two-level VSC. The three-level VSC described in Section 9.3.1.3 is considered the simplest example of this topology. Additional levels can be achieved by adding extra switches and diodes, however, the component count increases with the level number. An m-level DCC would require $(m - 1)$ capacitors and $(m - 1)(m - 2)$ clamping diodes. Each switch is only required to block a voltage level of $V_{dc}/(m - 1)$. The clamping diodes need to have unequal voltage ratings.

When the VSC is used for real power transfer, balancing of the capacitor voltages is one of the major challenges in this topology. Figure 9.13 shows a five-level DCC and its charging and discharging wave shapes under unity and zero power factor operations. Under unity power factor, at each half cycle the capacitors will be charged for an uneven period as shown in Figure 9.13. This results in an imbalance of the capacitor voltages. In the case of zero power

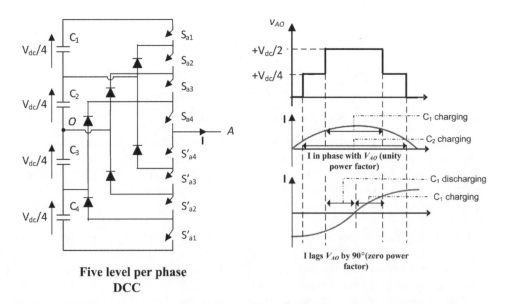

Figure 9.13   Charging and discharging of capacitors in five-level DCC

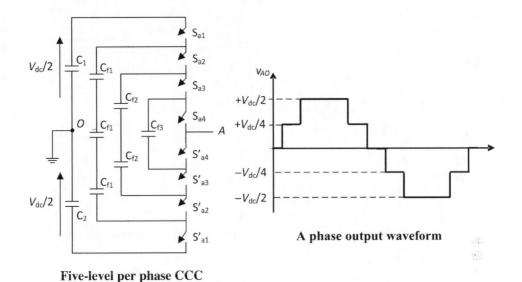

Five-level per phase CCC   **A phase output waveform**

Figure 9.14 Single-phase circuit and output of the five-level CCC configuration

factor operation, at each half cycle a capacitor is charged and discharged an equal amount of time (only the upper capacitor discharging and charging is shown in Figure 9.13) and the capacitor unbalance problem does not arise. The voltage imbalance problem can be solved by connecting separately controlled DC sources across each capacitor, but this increases the circuit complexity and cost.

The capacitor-clamped multi-level converter employs a large number of capacitors to form the step-like output. Figure 9.14 shows a single-phase five-level Capacitor Clamped (sometimes known as a floating or flying capacitor) Converter (CCC) configuration. The main DC capacitors C_1 and C_2 are used by all three phases and are usually fed from separate energy sources. The other capacitors C_{f1}, C_{f2}, C_{f3} float with respect to earth, leading to the term floating capacitors. The CCC configuration has an identical structure to floating capacitors and series-connected switching devices are used for all three phases.

The phase output of the five-level CCC configuration with respect to the mid-point of the DC link is shown in Figure 9.14. The size of the output voltage steps depends on the capacitor voltages. The switch pairs (S_{a1}, S'_{a1}), (S_{a2}, S'_{a2}), (S_{a3}, S'_{a3}) and (S_{a4}, S'_{a4}) are switched in opposition with a small time delay for switching transitions. The capacitors can be connected in such a way that their voltages can be added or subtracted. The combination of the selected capacitors determines the DC levels of the output voltage.

As separate floating capacitors are used in each phases, the fundamental frequency current flows in them, thus requiring large capacitors to prevent a variation of fundamental frequency voltage.

The CCC has several switching combinations to produce a DC level of the output voltage. These combinations can be used to balance the capacitor voltages and to solve the unequal duty problems in the switches [18]. However, a higher switching frequency is needed to maintain the capacitor voltages.

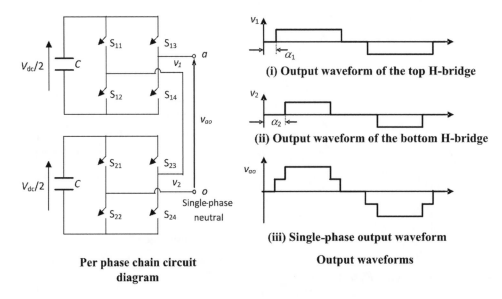

Figure 9.15   Single-phase circuit and output of the five-level M2C configuration

9.3.2.3   Multi-modular converters

A number of H-bridge converters can be connected to form a Multi-Modular Converter (M2C). Due to its modular nature, redundancy can be incorporated and 'standard' modules can be used.

Figure 9.15 shows a single-phase five-level multi-modular circuit. Sometimes the M2C configuration is called a chain circuit VSC. This arrangement has two identical H-bridge converter cells with separate DC sources, which allow independent control of the converter cells. Compared to the five-level FCC, the five-level M2C configuration requires only two DC capacitors.

Figure 9.15 shows the output voltages of the single-phase M2C circuit. In this configuration the number of levels can be defined as the number of DC voltages it can produce across the single-phase terminals. Each H-bridge can generate three voltage levels: when S_{12} and S_{13} are on, the voltage level is $+V_{dc}$, when S_{11} and S_{14} are on, the voltage level is $-V_{dc}$ and when all switches are off, the voltage level is zero. The switching angles α_1 and α_2 can be used to control the fundamental voltage component and eliminate one selected harmonic from the output voltage. Some system controls vary the DC-link voltages to control the fundamental voltage component. In that case, the switching angles α_1 and α_2 can be used to eliminate two selected harmonic components from the output voltage.

A dominant second harmonic current is caused by the full bridge single-phase converter arrangement, and this leads to a ripple on the DC link voltage. Therefore, the DC capacitors have to be oversized to maintain a reasonably low ripple DC voltage. In addition, the circuit also requires a complex control to maintain the DC voltage levels, particularly under system disturbances [18].

References

[1] Arrillaga, J. (1998) *High Voltage Direct Current Transmission*, IET, Stevenage.

[2] Kim, C., Sood, V.K. and Seong-Jo, G.J. (2010) *HVDC Transmission: Power Conversion Applications in Power Systems*, Wiley-IEEE, Singapore.

[3] Barker, C. *et al.* (2010) *HVDC: Connecting to the Future*, Alstom Grid.

[4] Song, Y.H. and Johns, A.T. (1999) *Flexible AC Transmission Systems*, IET Power and Energy Series 30.

[5] Acha, E., Fuerte-Esquivel, C.R., Ambriz-Pérez, H. and Angeles-Camacho, C. (2004) *FACTS: Modelling and Simulation in Power Networks*, John Wiley & Sons Ltd, Chichester.

[6] Jenkins, N., Ekanayake, J.B. and Strbac, G. (2010) *Distributed Generation*, IET, Stevenage.

[7] Heier, S. (2006) *Grid Integration of Wind Energy Conversion Systems*, John Wiley & Sons, Ltd, Chichester.

[8] Anaya-Lara, O., Jenkins, N., Ekanayake, J. *et al.* (2009) *Wind Energy Generation: Modelling and Control*, John Wiley & Sons, Ltd, Chichester.

[9] Hingorani, H. (1995) Introducing custom power. *IEEE Spectrum*, **32**(6), 41–48.

[10] Mohan, N., Undeland, T.M. and Robbins, W.P. (1995) *Power Electronics: Converters, Applications and Design*, John Wiley & Sons, Inc, New York.

[11] Yazdani, A. and Iravani, R. (2010) *Voltage-Sourced Converters in Power Systems*, IEEE-Wiley, Singapore.

[12] Holtz, J. (1992) Pulsewidth modulation – a survey. *IEEE Transactions on Industrial Electronics*, **39**(5), 410–420.

[13] Handley, P.G. and Boys, J.T. (1992) Practical real-time PWM modulators: an assessment. *IEE Proceedings-B (Electric Power Applications)*, **139**(2), 96–102.

[14] Shen, J. and Butterworth, N. (1997) Analysis and design of a three-level PWM converter system for railway-traction applications. *IEE Proceeding Electric Power Applications*, **144**(5), 357–371.

[15] Nabae, A., Takahashi, I. and Akagi, H. (1981) A new neutral-point-clamped PWM inverter. *IEEE Transactions on Industry Applications*, **1A-17**(5), 518–523.

[16] Withanage, R., Crookes, W. and Shammas, N. (2007) *Novel voltage balancing technique for series connection of IGBTs*. 2007 European Conference on Power Electronics and Applications, 2007, pp. 1–10.

[17] Hochgraf, C., Lasseter, R., Divan, D. and Lipo, T.A. (1994) *Comparison of multilevel inverters for Static VAr Compensation*. IEEE Industry Applications Society Annual Meeting, Vol. 2, 1994, pp. 921–928.

[18] Lai, J. and Peng, F.Z. (1996) Multilevel converters – a new breed of power converters. *IEEE Transactions on Industry Applications*, **32**(3), 509–516.

10

Power Electronics in the Smart Grid

10.1 Introduction

In the future Smart Grid there will be increasing connection to the distribution network of renewable energy sources, electric vehicles and heat pumps. More flexible loads will be expected to support the grid by accepting varying supplies of energy from renewable sources and by controlling peaks in demand. For sensitive loads such as computers and high value manufacturing plants, the quality of supply will be important. Therefore visibility, controllability, and flexibility will be essential features throughout the future power system with power electronics playing a key role

The power output of some of renewable energy sources is always DC (for example, photovoltaic systems) and an inverter is needed to interface them to the AC grid. Even though renewable energy sources using an AC generator (for example, wind turbines) can be connected directly to the grid, often some form of AC to DC and then DC to AC conversion is used. Some power system operating conditions demand rapid independent control of the active and reactive power output of the renewable energy generators. These control actions can only be achieved conveniently using a power electronic interface.

With the connection of a large number of distributed generators, including micro-generators, traditional methods of active power/frequency control and reactive power/voltage control will no longer be effective. The traditional voltage control method in a distribution circuit is an on-load tap changer and automatic voltage control relay, sometimes with line drop compensation. This control system may not operate satisfactorily when power flow in the distribution circuit becomes reversed. Furthermore, the tap changer is a mechanical device whose operating time may be not fast enough for some dynamic voltage control functions. In these circumstances, a fast acting reactive power source such as a distribution STATCOM is useful, particularly for inductive circuits.

Smart Grid: Technology and Applications, First Edition.
Janaka Ekanayake, Kithsiri Liyanage, Jianzhong Wu, Akihiko Yokoyama and Nick Jenkins.
© 2012 John Wiley & Sons, Ltd. Published 2012 by John Wiley & Sons, Ltd.

10.2   Renewable energy generation

Renewable energy sources are being developed in many countries to reduce CO_2 emissions and provide sustainable electrical power. The balance of particular technologies and their scale changes from country to country. However, hydro, wind, biomass (solid biomass, bioliquids and biogas), tidal stream, and photovoltaic (PV) are common choices.

Variable speed turbines are used for wind, small hydro and tidal power generation. These generally use AC–DC–AC power conversion where the turbine is arranged to rotate at optimum speed to extract the maximum power from the fluid flow or minimise mechanical loads on the turbine. The variable frequency power output from the generator is first converted to DC. A second converter is used to convert DC into 50/60 Hz AC.

The output of a PV system is DC and therefore a DC–AC converter is essential for grid connection. Biomass technologies use a steam or gas turbine and a conventional generator. Reciprocating engines may be fuelled by biogas. They generally use a synchronous generator connected to the grid directly and are not considered in this chapter.

The power electronic interface between a renewable energy source and the grid can be used to control reactive power output and hence the network voltage as well as curtailing real power output, and so enable the generator to respond to the requirements of the grid.

10.2.1   Photovoltaic systems

Photovoltaic (PV) systems which convert solar power directly into electricity are being installed in increasing numbers in many countries, for example, Germany, Spain, the USA and Japan. Feed-in-tariffs, which provide guaranteed payment per unit of electricity (p/kWh) for renewable electricity generation have been particularly important in stimulating the uptake of PV.

Figure 10.1 shows the main elements of a grid-connected domestic PV system. It typically consists of: (1) a DC–DC converter for Maximum Power Point Tracking (MPPT) and to increase the voltage; (2) a single phase DC–AC inverter; (3) an output filter and sometimes a transformer; and (4) a controller.

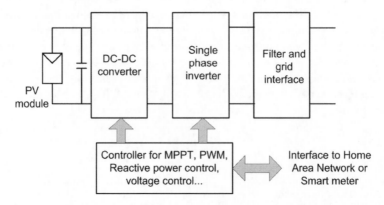

Figure 10.1   Block diagram of a domestic PV system

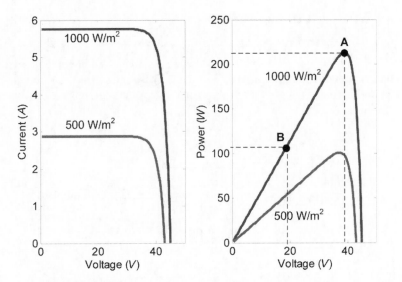

Figure 10.2 Typical current/voltage and power/voltage characteristics of a PV module for irradiance of 1000 W/m^2 and 500 W/m^2

The PV module contains a number of photovoltaic cells connected in series and in parallel. Figure 10.2 shows the current versus voltage and the power versus voltage characteristics [1] of a PV module. The maximum power output of the module is obtained near the knee of its voltage/current characteristic.

Different configurations of DC–DC converters are used, for example, boost, push–pull, full bridge, and flyback converter [2, 3]. The DC voltage on the inverter side of the DC–DC converter is normally maintained to be constant by the inverter control. The MPPT algorithm is used to find continually a PV array DC voltage which extracts the most power from the PV array while the cell temperatures and operating conditions of the module change.

As it is easy to implement in a digital controller, the most widely used MPPT algorithm is 'perturb and observe' sometime known as 'hill climbing'. In this method, the terminal voltage of the PV array is perturbed in one direction and if the power from the PV array increases, then the operating voltage is further perturbed in the same direction. Otherwise if the power from the PV array decreases, then the operating voltage is perturbed in the reverse direction. Another technique more easily implemented with analogue electronics is incremental conductance. This is based on the fact that at maximum power point, $(di/dv) + (i/v)$ of the PV array is zero (derived from $dP/dv = 0$) [4]. This equation suggests that the voltage corresponding to the maximum power can be found by measuring the incremental conductance (di/dv) and instantaneous conductance (i/v).

The DC voltage obtained from the DC–DC converter is inverted to 50/60 Hz AC. A voltage source inverter is widely used. As discussed in Chapter 9, this normally uses a pulse width modulation switching technique to minimise harmonic distortion. Finally, a filter is placed at the output to minimise harmonics fed into the power system. In some designs a transformer is also employed at the output of the inverter to ensure no DC is injected into the grid.

Example 10.1

The PV system shown in Figure 10.3 has two series-connected PV modules with the V-I characteristic shown in Figure 10.2. The single phase inverter operates with sinusoidal PWM and is connected directly to the 230 V mains. The irradiance on the module is 1000 W/m^2.

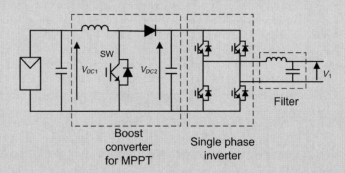

Figure 10.3   Figure for Example 10.1

1. Describe a possible control strategy which could be used.
2. What should be the amplitude modulation index of the inverter to maintain V$_{DC2}$ at 350 V?
3. Calculate the duty ratio of the switch SW that is required to extract maximum power.
4. If, due to constraints of the local power network, the output of the PV system was reduced by 50 per cent, calculate the new duty ratio required for switch SW.

Data: For the boost converter: $V_{DC2} = (1/(1-D)) \times V_{DC1}$ where D is the duty ratio of the switch SW.

For a single phase inverter operating with sinusoidal PWM $V_1 = m_a \times V_{DC2}$ where V_1 is the peak value of fundamental of the inverter output voltage, m_a is the modulation index [2].

Answer

1. V_{DC2} is always maintained at a reference value by changing the modulation index of the inverter. When the MPPT algorithm detects that V_{DC1} should be increased, then the duty ratio, D, is reduced. On the other hand when the MPPT algorithm detects that V_{DC1} should be decreased, then the duty ratio, D, is increased.
2. V_1 is $230 \times \sqrt{2} = 325$ V. Hence from $V_1 = m_a \times V_{DC2}$, to maintain V_{DC2} at 350 V:

$$m_a = \frac{325}{350} = 0.93.$$

3. In order to extract peak power, voltage V_{DC1} should be maintained at 80 V (voltage across each series modules should be maintained approximately at 40 V as shown in point A on Figure 10.2).

Then:
$$350 = \frac{1}{1-D} \times 80$$

Therefore, $D = 0.77$.

4. If the power output of the PV system needs to be reduced by 50 per cent, as the irradiance is not changed, V_{DC1} should be changed to 40 V (see point B on Figure 10.2).

Then:
$$350 = \frac{1}{1-D} \times 40$$

Therefore, $D = 0.89$

10.2.2   Wind, hydro and tidal energy systems

Wind, hydro and tidal generation systems all involve converting the potential and/or kinetic energy in water or air into electrical energy. In recent years there has been a dramatic increase in power generation from the wind with the capacity of wind turbines that have been installed across the globe now approaching 200 GW. Hydropower is a mature technology with units varying in size from a few kW to hundreds of MW. Tidal stream generation is a more recent innovation and the subject of considerable research and development effort.

Wind farms are now being developed both onshore and offshore. Placing a wind turbine in the sea is more challenging and expensive but offshore wind farms enjoy a stronger and more consistent wind resource and reduced environmental impact. The majority of generators used in offshore wind turbines are variable speed. Some years ago, fixed speed wind turbines were common onshore but the majority of new onshore installations also use variable speed wind turbines.

In developed countries the economically attractive sites for large hydropower plants have now almost all been exploited but there are still opportunities for the development of small and micro hydropower plants. The efficiency of these smaller units across a wide range of water flows can be improved by using a variable speed generator with a power electronic interface.

Different turbine designs are available for tidal stream technologies [5]. Some stand on the seabed and are shrouded and some are floating. They can be categorised into:

1. *Horizontal axis turbines*: The architecture of these turbines is similar to that of wind turbines.
2. *Vertical axis turbines*: Vertical axis turbines do not require orientation into the flow but do suffer from large cyclic torques.
3. *Oscillating hydrofoil devices*: These have hydrofoils which move back and forth in a plane normal to the tidal stream.
4. *Ducted devices*: The tidal flow is directed through a duct and a smaller diameter turbine is situated inside the duct.

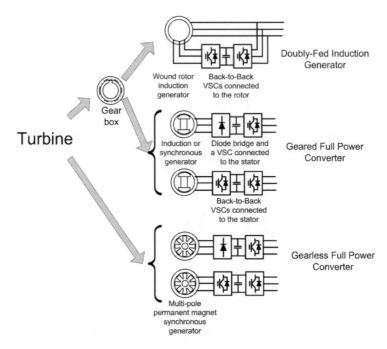

Figure 10.4   Different variable speed generator configurations

10.2.2.1   Power electronic converters

For variable speed operation of wind, hydro and tidal stream turbines, Doubly Fed Induction Generators (DFIG) or Full Power Converter (FPC) based generators (see Figure 10.4) may be used. The DFIG has a wound rotor induction machine where the rotor is connected to back-to-back power electronic converters. The four quadrant converters control both active and reactive power flow to and from the rotor circuit. The rotor speed can be changed by absorbing or injecting active power by the rotor side converter.

In the FPC generator, a diode bridge or a four-quadrant converter is connected to the stator terminals of the generator. The generator may be a synchronous or cage asynchronous machine. The generator speed is controlled to track the maximum power and the generator output frequency varies with wind/hydraulic/tidal flow conditions. The variable frequency power is then converted to DC using the power electronic converter and then inverted back to 50 Hz AC. The FPC configuration also allows operation without a gearbox. A multi-pole permanent magnet machine with a large number of poles (as high as 100) is used in gearless designs.

10.2.2.2   Control of wind turbines

A variable speed wind turbine uses back-to-back Voltage Source Converters (VSC) to control the rotor speed so that it can maintain maximum aerodynamic efficiency at varying wind speeds. Figure 10.5 shows the power output of a 75 m diameter rotor against generator rotor speed with different wind speeds. Below rated wind speed, the generator speed is varied using a power electronic controller to track the maximum power curve. This is normally achieved using a torque controller where a torque set point, corresponding to the maximum power curve, is stored in a look-up table with respect to rotational speed. The actual electromechanical torque developed by the converter is compared against it. When the rated rotor speed is reached at around 12 m/s wind speed, the blade pitch is altered to reduce the power generated by the aerodynamic rotor (point X). Below point X only the torque controller is active and pitch demand is fixed at an optimum value, normally $-2°$. Above the rated wind speed, the pitch controller is active, and the electronic controller maintains set point torque at its rated value.

Variable speed wind turbines can be controlled to support the grid [6, 7, 8]. For example, when a wind turbine operates on electronic control through its back-to-back converters, the power extracted from the wind can be reduced by changing the operating speed of the generator. This is useful when a high frequency occurs on the grid network, perhaps due to the sudden disconnection of a large load. Wind turbines can also provide low frequency response, if they are de-loaded during normal operation.

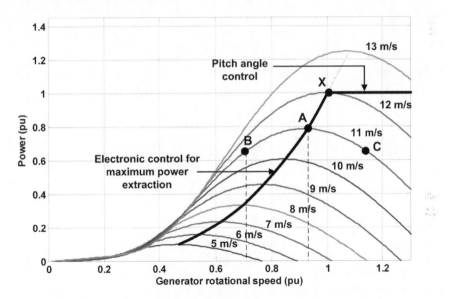

Figure 10.5 Variable speed wind turbine operating regions. *Note:* Power base is 2 MW and speed base is 1800 rev/min

Example 10.2

A wind turbine having the characteristic of Figure 10.5 operates in a wind speed of 11 m/s. The generator used is a synchronous generator. Due to a high frequency on the network, the grid operator requests a reduction of the wind turbine output by 20 per cent. At rated speed the frequency modulation index of the wind turbine side VSC is 30. What should be the new modulation index required to reduce the wind turbine power output?

Answer

At 11 m/s, the wind turbine output is given by the interception of the maximum power extraction curve and 11 m/s curve. Thus, the turbine operates at point A (Power $\approx$ 0.8 pu) of Figure 10.5. When power is reduced by 20 per cent, the new operating point is B (Power $= 0.8 \times 0.8 = 0.64$ pu).

In order to obtain this power, the speed of the turbine could either be reduced to point B or increased to point C. Operation at point C is limited by the maximum speed of the turbine.

The corresponding speed for operation at point B is approximately 0.7 pu, that is: $1800 \times 0.7 = 1260$ rev/min.

For a synchronous machine, the speed is proportional to frequency at the stator (say, proportional constant k). At 1800 rev/min, the frequency of the triangular wave carrier is:

$$f_c = \frac{1800}{k} \times 30 \tag{10.1}$$

Since the carrier frequency is the same, to obtain a speed of 1260 rev/min, the frequency modulation index, m_f, should satisfy:

$$f_c = \frac{1260}{k} \times m_f \tag{10.2}$$

From Equations (10.1) and (10.2): $m_f = 43$

10.2.2.3 Control of hydro turbines

The design of a hydro turbine is optimised for a defined rotational speed, hydraulic head and discharge. As the hydraulic conditions change, the conversion efficiency of the turbine also changes. Variable speed operation changes the turbine speed so as to maximise its efficiency over a range of different hydraulic conditions [9, 10].

The net head and the opening of the guide vanes control the discharge through the turbine. The turbine performance is represented by the so-called hill chart. The hill chart gives the efficiency curve of the turbine for different flow rates as the rotor speed changes. As can be seen from Figure 10.6, for a given discharge, there is an optimum rotor speed which gives the maximum efficiency (see solid dotted line). Once the hill chart has been established from model tests, then it can be used for speed control of the generator by using a look-up table.

10.2.2.4 Control of tidal stream turbines

In tidal stream devices, in order to extract maximum power, the torque presented by the generator to the prime mover varies with the tidal flow conditions. One possible control approach is to use a hill climbing technique, as described for PV systems. Alternatively, a control concept similar to that used for wind turbines may be employed for tidal energy

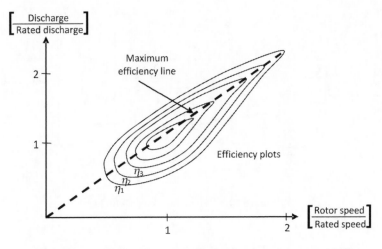

Figure 10.6   Hill chart for a propeller turbine ($\eta_1 < \eta_2 < \eta_3 \ldots$) [10]

converters where the power extracted is determined by off-line calculations of the rotor angular velocity relative to the tidal stream flow and the creation of a look-up table.

10.3   Fault current limiting

It is anticipated that more distributed generation will be connected to the future Smart Grid power system. Some of these generators will be the synchronous type, giving a sustained current into a short circuit fault. Some of them will be asynchronous (induction) where the symmetrical fault current decays within less than a second. There will also be a large number of distributed generators that are connected through power electronic converters, whose fault current contribution is limited by the rating of the electronic switches they employ and their control system. The connection of different types of distribution generation will introduce issues such as:

1. In some circuits, the synchronous generators will increase the fault current through switchgear (which has a limited interruption capability), thus requiring replacement of the switchgear or other measures to limit the short circuit current.
2. In some other circuits, which are rich in power electronic connected distributed generators, the fault current contribution may not be adequate to ensure detection of the fault by existing over-current protection, thus demanding different protection methods.

Fault Current Limiters (FCL) in distribution circuits limit the fault current. Different designs such as current limiting fuses, superconducting, magnetic, static and hybrid fault current limiters exist. However, only current limiting fuses are widely used at present [11].

A number of different static FCL designs are reported in the literature [12, 13, 14]. Figure 10.7 shows three of them, each using a different kind of switch. The only design that has been implemented in a real system is the scheme shown in Figure 10.7a. The series compensator in the Kayenta substation in the USA includes this capability [12]. In this design,

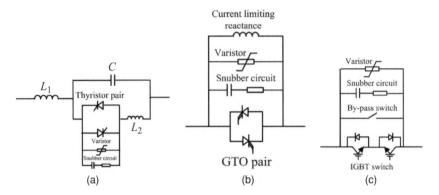

Figure 10.7   Static fault current limiters

under normal operation, both thyristors are off and current flows through L_1 and C. The impedance of L_1 and C is selected such that at 50/60 Hz, $\omega L_1 - 1/\omega C = 0$. This introduces a zero impedance in series with the line. When there is a fault, both thyristors go into full conduction mode and L_2 is connected in parallel with C, thus increasing the total inductive impedance in series with the line. A varistor is connected to limit the transient voltage across the thyristors. The rate of rise of the transient voltage is limited by the snubber circuit.

In the design shown in Figure 10.7b, back-to-back Gate Turn Off (GTO) thyristors are placed in series with the line. When a fault is detected, the normally conducting GTOs are switched off and the current is diverted to the parallel reactance. This limits the fault current. In [13], this circuit is proposed for both fault current limiting and interruption. It is possible to continuously switch on and off the GTO switches so that the fault current is shared between the GTOs and the parallel connected reactor.

In the design shown in Figure 10.7c, under normal operation, the bypass switch takes the current. Whenever there is a fault, the bypass switch is opened and the IGBTs take over the current flow. The IGBTs are switched off when the fault current reaches a pre-set value. Then the fault current flows through the varistor. The clamping voltage of the varistor is set to be higher than the peak supply voltage. Therefore, the fault current starts decreasing. When the current reaches a minimum value, the IGBT is turned on again. In this way the fault current is maintained below a maximum value.

Example 10.3

For the part of the distribution network shown in Figure 10.8, calculate the following:

1. The fault current for a three-phase fault at E without generator G1.
2. The fault current for a three-phase fault at E with generator G1 (without Static FCL).
3. The values of C and L_2 to reduce the fault current calculated in (2) by 5 per cent.

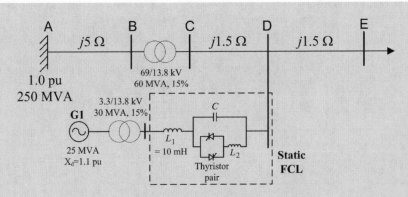

Figure 10.8   Figure for Example 10.3

Answer

Select $S_{base} = 250$ MVA

At 69 kV, $Z_{base} = \dfrac{\left(69 \times 10^3\right)^2}{250 \times 10^6} = 19.04\ \Omega$ and

$$I_{base} = \frac{250 \times 10^6}{\sqrt{3} \times 69 \times 10^3} = 2091.85\ \text{A}$$

At 13.8 kV, $Z_{base} = \dfrac{\left(13.8 \times 10^3\right)^2}{250 \times 10^6} = 0.76\ \Omega$ and

$$I_{base} = \frac{250 \times 10^6}{\sqrt{3} \times 13.8 \times 10^3} = 10459.25\ \text{A}$$

Per unit impedances are given in Table 10.1.

Table 10.1   Per unit impedances

Line AB	Transformer 69/13.8 kV	Line CD	Line DE	G1	Gen Transformer
$j5/19.04$ $= j0.26$	$j0.15 \times 250/60$ $= j0.625$	$j1.5/0.76$ $= j1.97$	$j1.5/0.76$ $= j1.97$	$j1.1 \times 250/25$ $= j11$	$j0.15 \times 250/30$ $= j1.25$

1. Fault current without generator G1

$$= \frac{1}{0.26 + 0.625 + 1.97 + 1.97} = 0.207\ \text{pu}$$

$$= 0.207 \times 10459.25 = 2166.8\ A$$

2. The equivalent circuit with generator G1 is given by:

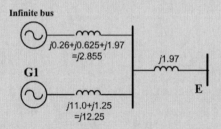

Figure 10.9 Equivalent circuit with generator G1

$$\text{Fault current with generator G1} = \frac{1}{2.855//12.25 + 1.97} = 0.233 \text{ pu}$$

$$= 0.233 \times 10459.25 = 2440.7\,A$$

3. If the impedance of the fault current limiter is jZ_{FCL}, then the equivalent circuit is as shown in Figure 10.9.

If the fault current is reduced by 5 per cent, the new fault current $= 0.233 \times 0.95 = 0.22$ pu

$$\text{Therefore, } \frac{1}{\left[\dfrac{2.855 \times (12.25 + Z_{FCL})}{(2.855 + 12.25 + Z_{FCL})} + 1.97\right]} = 0.22 \text{ pu}$$

Solving this equation, it can be found that $Z_{FCL} = j14.03$ pu
From the FCL circuit shown in Figure 10.10.

$$100\pi L_1 - 1/100\pi C = 0$$

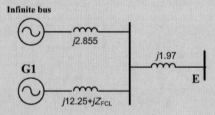

Figure 10.10 Equivalent circuit with FCL

Since $L_1 = 10\,\text{mH} \rightarrow C = 1013\,\mu\text{F}$

$$jZ_{FCL} = j\omega L_1 + \frac{j\omega L_2 \times 1/j\omega C}{j\omega L_2 + 1\,j\omega C}$$

Since $\omega L_1 - 1/j\omega C = 0$, therefore $1/j\omega C = -j\omega L_1$, by substituting this in the previous equation, the following equation can be obtained:

$$Z_{FCL} = \omega L_1 + \frac{\omega L_2 \times -j\omega L_1}{j\omega L_2 - j\omega L_1}$$

$$= \omega L_1 \left[1 + \frac{L_2}{L_1 - L_2}\right]$$

Substituting for Z_{FCL} and L_1:

$$14.03 = 100\pi \times 10 \times 10^{-3} \left[1 + \frac{L_2}{10 \times 10^{-3} - L_2}\right]$$

Solving this equation $L_2 = 7.76$ mH.

The network shown in Figure 10.3 was simulated using IPSA software and the fault current was obtained with and without the FCL and shown in Figure 10.11. For the simulations the resistance of lines AB (2.5 Ω), CD (1.5 Ω) and DE (1.5 Ω) were considered (these values were ignored in the worked example to simplify the calculations).

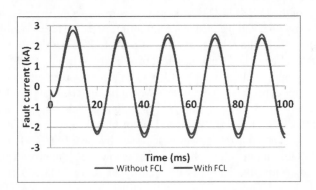

Figure 10.11 Fault current from the generator with and without FCL

10.4 Shunt compensation

For many years shunt capacitors or reactors have been connected to the power system, sometimes with thyristor control, in order to manage reactive power. More recently, shunt compensation devices based on voltage source inverters such as STATCOMs, active filters and Voltage Source Converters with Energy Storage (VSC-ES) have begun to be used in the power system. STATCOMs are used to provide reactive power compensation in both transmission and distribution circuits in order to manage network voltages, reduce losses and overcome possible instabilities.

The voltage change across a distribution circuit is given by

$$\Delta V = (PR + QX)/V \tag{10.3}$$

where P and Q are active and reactive power flows, X and R are the reactance and resistance of the circuit and V is the nominal voltage.

As the power output of a distributed generator driven by a renewable energy source varies, the voltage change across the circuit to which it is connected also varies. Mitigating these voltage fluctuations can be effected with shunt compensation devices such as a STATCOM or a VSC-ES that can vary Q or P and Q with the change of voltage.

Many electronic loads draw non-sinusoidal currents and this can lead to unacceptable levels of voltage distortion. Standards such as IEEE 519–1992 [15] and ER G5/4 [16] specify limits to either the harmonic current which may be injected or to the resulting harmonic voltages on the network. If these limits are exceeded, then connection to the network is refused. If the load equipment cannot be modified economically to approach the sinusoidal current, then power electronic active filters may be used to correct the current drawn from the network.

10.4.1 D-STATCOM

The STATCOM is the power electronic counterpart of the traditional rotating synchronous condenser. A STATCOM connected to the distribution circuits is normally called a D-STATCOM. The basic operating principle of a D-STATCOM for reactive power control is shown in Figure 10.12. The VSC of the D-STATCOM produces a controllable three-phase voltage ($V_{STATCOM}$) in phase with the terminal voltage. When the amplitude of the VSC output voltage ($V_{STATCOM\_1}$) is less than the terminal voltage ($V_{terminal}$), the D-STATCOM draws a current ($I_{STATCOM\_1}$) lagging the terminal voltage, thus absorbing reactive power. When the VSC generates a voltage higher than the terminal voltage, the D-STATCOM generates reactive power.

Two main control approaches for STATCOM control can be found in the literature. One is phase shifting control [17, 18]. When $V_{STATCOM}$ slightly leads $V_{terminal}$, net real power flows from the D-STATCOM to the AC system. This in turn decreases the DC capacitor voltage and thus $V_{STATCOM}$. Then reactive power is absorbed by the STATCOM. The converse occurs when the STATCOM output voltage slightly lags below the system voltage, the DC capacitor voltage rises and so reactive power is injected by the STATCOM. Phase shifting does not

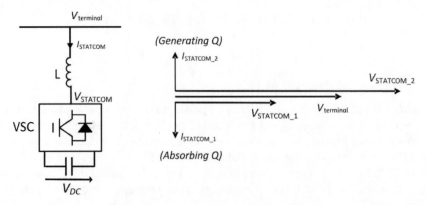

Figure 10.12   Operation of a D-STATCOM

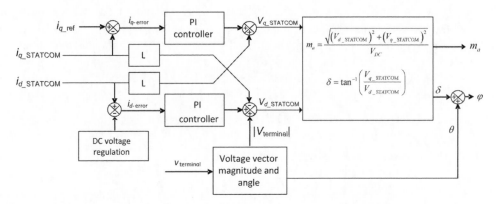

Figure 10.13   A decoupled current control method for D-STATCOM [18]

rely on PWM control of the voltage source inverter and so is attractive for larger STATCOMs where other modulation strategies are used to reduce the switching losses of the inverter.

Another control approach using PWM is shown in Figure 10.13. This regulates the modulation index (m_a) and phase angle (φ) of the inverter. The d-axis and q-axis vector components of the injected current (I_{STATCOM}) are calculated by taking the d-axis aligned with the terminal voltage vector (V_{terminal}). The current component I_{d_STATCOM} controls the DC-link capacitor voltage and the regulation of I_{q_STATCOM} provides the reactive power that should be injected or absorbed by the D-STATCOM.

Example 10.4

Consider the D-STATCOM shown in Figure 10.14. Assume that a six-pulse VSC operating on sinusoidal PWM is employed for the D-STATCOM. Calculate the modulation index required: (1) to generate 5 Mvar of reactive power; and (2) to absorb 5 Mvar of reactive power.

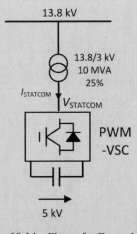

Figure 10.14   Figure for Example 10.4

Data: For a six-pulse VSC operating on sinusoidal PWM $V_{LL} = 0.612 \times m_a \times V_{DC}$ where V_{LL} is the line to line voltage at D-STATCOM terminals, m_a is the modulation index and V_{DC} is the DC capacitor voltage [2].

Answer

Assuming the base MVA is 10 MVA, the base primary voltage of the transformer is 13.8 kV and the base secondary voltage is 3 kV.

The transformer leakage reactance $= 0.25$ pu.
D-STATCOM terminal voltage $= 0.612 \times m_a \times 5/3 = 1.02 m_a$ pu.
The pu equivalent circuit is shown in Figure 10.15.

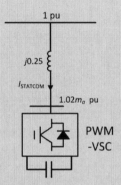

Figure 10.15   Equivalent circuit of Figure 10.14

1. When D-STATCOM is generating 5 Mvar, in pu $Q = +0.5$ pu
   Then from the pu equivalent circuit:

$$Q = 0.5 = \frac{1.02 m_a - 1}{0.25}$$

$$\therefore m_a = 1.1$$

When $m_a > 1$, the PWM-VSC operates in the over-modulation region and the linear relationship between m_a and STATCOM terminal voltage is no longer valid. To generate 5 Mvar of reactive power, m_a should be greater than 1.1, finding the exact value of m_a needs more precise calculation of the PWM output voltage.
2. When the D-STATCOM absorbs 5 Mvar, in pu $Q = -0.5$ pu
   Then from the pu equivalent circuit:

$$Q = -0.5 = \frac{1.02 m_a - 1}{0.25}$$

$$\therefore m_a = 0.86$$

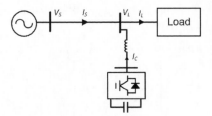

Figure 10.16   Load compensation

10.4.1.1   Load compensation

A D-STATCOM can be used for power factor correction and to balance the current drawn by an unbalanced load. A typical connection arrangement is shown in Figure 10.16. For load compensation, first, the current that the D-STATCOM should inject is calculated. Then a suitable controller generates this current. For simplicity, here it has been assumed that the D-STATCOM is an ideal current source. However, in practice, due to the limitations of the VSC and its controller as well as the active power required to charge the capacitor of the D-STATCOM, ideal compensation may not be possible.

Two compensation techniques are discussed here:

1. Power factor correction of a balanced three-phase load

   For unity power factor operation of the load, the D-STATCOM should inject a current to compensate for load reactive power, q_L.

   The instantaneous reactive power that the D-STATCOM should inject is given in [19] as:

$$v_{dL} \times i_{qC} - v_{qL} \times i_{dC} = q_L \tag{10.4}$$

   Here the d-axis is chosen to align with V_L.

v_{dL} and v_{qL} are the d and q axis component of V_L.

i_{dC} and i_{qC} and are the d and q axis component I_C.

   Assuming the D-STATCOM is not absorbing or injecting any real power:

$$v_{dL} \times i_{dC} + v_{qL} \times i_{qC} = 0 \tag{10.5}$$

   From Equations (10.4) and (10.5), the current that should be injected by the D-STATCOM can be calculated. A current control technique can then be applied to control the D-STATCOM to obtain the required currents at its output.

2. Power factor correction and balancing of an unbalanced three-phase load

Assuming the supply currents and voltages are balanced, instantaneous active power, p_S, and reactive power, q_S, are given by [20]:

$$p_S = v_{aS} \times i_{aS} + v_{bS} \times i_{bS} + v_{cS} \times i_{cS} \tag{10.6}$$

$$q_S = \begin{bmatrix} q_{aS} \\ q_{bS} \\ q_{cS} \end{bmatrix} = \begin{bmatrix} v_{bS} \times i_{cS} - v_{cS} \times i_{bS} \\ v_{cS} \times i_{aS} - v_{aS} \times i_{cS} \\ v_{aS} \times i_{bS} - v_{bS} \times i_{aS} \end{bmatrix} \tag{10.7}$$

v_{aS}, v_{bS} and v_{cS} are the three-phase voltages of V_S.

i_{aS}, i_{bS} and i_{cS} are the three-phase currents of I_S.

Assuming that load active power is equal to p_S and q_S is zero (assuming load reactive power is supplied by the D-STATCOM), from Equations (10.6) and (10.7), three currents i_{aS}, i_{bS} and i_{cS} can be found. Then the current that should be injected by the D-STATCOM is given by:

$$q = \begin{bmatrix} i_{aC} \\ i_{bC} \\ i_{cC} \end{bmatrix} = \begin{bmatrix} i_{aS} - i_{aL} \\ i_{bS} - i_{bL} \\ i_{cS} - i_{cL} \end{bmatrix} \tag{10.8}$$

In Equation (10.8), the three-phase load currents are not equal and the three phases of the D-STATCOM output should be controlled independently. One possible circuit is three H-bridges (see Figure 9.2) connected to a common DC capacitor as shown in Figure 10.17.

10.4.1.2 Voltage control

A control system that is used with thyristor-based static Var compensators may be used with a D-STATCOM for voltage control. Conventionally, voltage control with a current droop as shown in Figure 10.18 is utilised. A PI controller is used to determine the var demand signal by regulating the error signal between the actual voltage and a reference, V^*_{Ref}. The reference voltage is calculated by introducing a regulator droop (k) to the reference voltage (V_{ref}). The regulator droop allows the terminal voltage to be smaller than the nominal value at capacitive compensation and higher at inductive compensation.

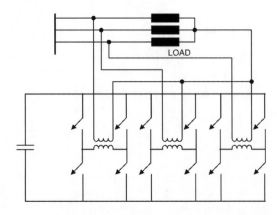

Figure 10.17   D-STATCOM with three H-bridges to give independent control of each phase

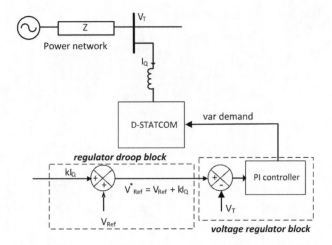

Figure 10.18   Closed loop control for voltage regulation [21]

In a case where the voltage V_T changes due to a variability of renewable energy source connected to that busbar, the D-STATCOM could supply or absorb reactive power to minimise fluctuations in the voltage [22].

Example 10.5

A 5 MW induction generator of a hydro scheme is connected to the distribution system as shown in Figure 10.19. When the induction generator generates 5 MW, it absorbs 2.5 Mvar of reactive power.

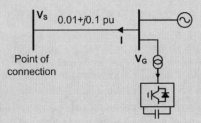

Figure 10.19   Figure for Example 10.5. *Note:* pu values are given on 10 MVA basis.

1. Write an expression for the voltage at the point of connection, $\mathbf{V_S}$, in terms of active power generated, P, net reactive power absorbed, Q, and generator terminal voltage, $\mathbf{V_G}$.
2. When the reactive power generation of the D-STATCOM is Q_C, it was found that the voltage at the point of connection is $1.0013\angle2.9°$ pu and the terminal voltage of the generator is $1\angle0°$ pu. Calculate the value of Q_C.

Answer

1.

$$\mathbf{V_G}\mathbf{I}^* = P - jQ$$

$$\mathbf{I} = \frac{P + jQ}{\mathbf{V_G^*}}$$

$$\mathbf{V_S} = \mathbf{V_G} + \left(\frac{P + jQ}{\mathbf{V_G^*}}\right)(0.01 + j0.1)$$

$$= \left(\mathbf{V_G} + \frac{0.01P - 0.1Q}{\mathbf{V_G^*}}\right) + j\left(\frac{0.01Q + 0.1P}{\mathbf{V_G^*}}\right)$$

2. When the D-STATCOM generates Q_C;

$$P = 5 \text{ MW} = 0.5 \text{ pu}$$
$$Q = 2.5 - Q_C \text{ Mvar}$$

With $\mathbf{V_S} = 1.0013\angle 2.9° = 1.0 + j0.0505$ pu and $\mathbf{V_G} = 1.0\angle 0° = 1$ pu

$$1.0 + j0.0505 = \left(1 + \frac{0.01 \times 0.5 - 0.1 \times Q}{1}\right) + j\left(\frac{0.01 \times Q + 0.1 \times 0.5}{1}\right)$$

$$= 1.005 - 0.1Q + j(0.05 + 0.01Q)$$

Equating the real or imaginary part of the above equation, it can be found that $Q = 0.05$ pu.

Therefore the reactive power generated by the D-STATCOM $= 2.5 - 10 \times 0.05 = 2.0$ Mvar.

10.4.2 Active filtering

A shunt active filter consists of a PWM-controlled current or voltage source converter. It injects the harmonic currents absorbed by the non-linear load such as an arc furnace, a rectifier-fed load (motor, heater) or a thyristor control motor drive. Figure 10.20 shows a voltage-controlled shunt active filter. The harmonic current drawn by a thyristor-controlled DC motor drive load is shown in the second plot of Figure 10.21. If the active filter supplies the harmonic current shown in the fourth plot, then the resultant current drawn from the source is sinusoidal.

10.4.3 Shunt compensator with energy storage

With recent advances in energy storage technology, the application of a VSC-ES has now become a feasible option for steady state voltage control and elimination of power system disturbances [23, 24].

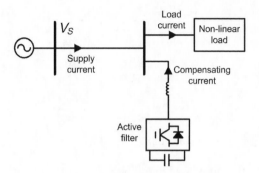

Figure 10.20   Shunt active filter

The VSC-ES can be controlled to exchange both real and reactive power with the AC system. The real and reactive power can be controlled independently of each other and any combination of real power generation/absorption with reactive power generation/absorption is achievable (within the ratings of the equipment).

As with a D-STATCOM, the reactive power generation or absorption capability of the VSC-ES can be used for load compensation or steady state and transient voltage control. The active power generation or absorption capability of the VSC-ES can be used to enhance steady state and transient voltage control, to eliminate voltage sags and to damp power oscillations.

Figure 10.22 shows a typical VSC-ES connection. In this generic example, even though *P* and *Q* are marked for a load, it could be for a DG (in that case, *P* will reverse and the direction

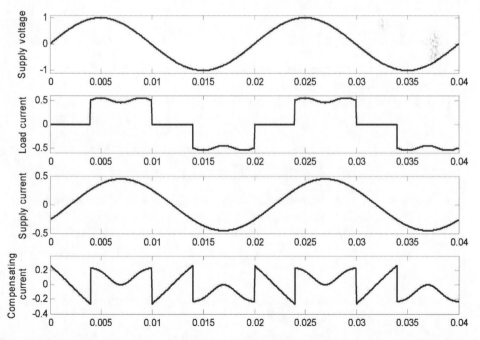

Figure 10.21   Load and compensating current for a thyristor control DC motor drive. *Note:* All the values in the y axis are in pu

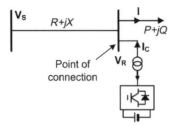

Figure 10.22   VSC-ES connection

of Q will be determined by the type of machine employed for the DG). The following equations are derived for this circuit (phasors are represented in bold):

$$\mathbf{I} = \frac{P - jQ}{\mathbf{V_R^*}} \tag{10.9}$$

$$\frac{\mathbf{V_S} - \mathbf{V_R}}{R + jX} + \mathbf{I_C} = \mathbf{I}$$

$$\mathbf{V_R} = \mathbf{V_S} - (\mathbf{I} - \mathbf{I_C})\,(R + jX) \tag{10.10}$$

Without the VSC-ES, the voltage $\mathbf{V_R}$ varies with the variation of P and Q. It is possible to find the current that should be injected by the VSC-ES, to minimise or eliminate variation of $\mathbf{V_R}$. Once the $\mathbf{I_C}$ that should be injected is known, then a current control technique can be used to provide a compensation current.

Example 10.6

A 50 MW wind farm is connected to the distribution system. Due to the blades passing the towers, it was observed that the power generated by the wind farm is given by: $40 + 4\sin(10t)$ MW. Calculate the current that should be injected by the VSC-ES to compensate for the blade-passing frequency of the power. For simplified calculations, assume that reactive power at the point of connection is zero. Further assume that with the VSC-ES, the voltage at the point of connection is constant at 33 kV.

Answer

From Equations (10.9) and (10.10), in order to eliminate voltage oscillations at the point of connection, the VSC-ES should inject a time-varying component of the power.

The current corresponding to the time varying component of the power is:

$$\mathbf{I} = \frac{4 \times 10^6 \times \sin(10t)}{33 \times 10^3/\sqrt{3}} = 210\,\sin(10t)\,\text{A}$$

If the VSC-ES injects the same amount of time-varying current, then the voltage at the point of connection will be constant.

So, $\mathbf{I_C} = 210\,\sin(10t)\,\text{A}$

The applications of VSC-ES include:

1. *Load compensation*: In many applications, the VSC-ES is operated as a load compensator where it fully supplies the load reactive power requirement in the steady state, thus maintaining the load power factor near unity.
2. *Steady state voltage control*: The VSC-ES can be used for steady state voltage control plus mitigation of other system disturbances. Under heavily loaded conditions, the VSC-ES supplies P and Q to mitigate voltage variations. On the other hand, when the system is lightly loaded, battery banks in the VSC-ES can be charged.
3. *Sag mitigation*: The application of a shunt device for mitigation of voltage sags has added advantages when compared to that of a series device, as the shunt devices can simultaneously be used for steady state voltage control, power oscillation damping and as a back-up power source. However, as the shunt device can compensate only for a small voltage dip when compared to a series device, the effectiveness for sag mitigation is limited. Figure 10.23 shows some experimental results obtained to demonstrate sag mitigation capability of the VSC-ES.

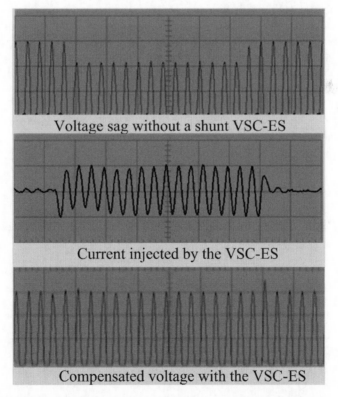

Figure 10.23 Experimental result showing sag minimisation of using a VSC-ES. *Note:* For clarity only the positive half of the voltage is shown

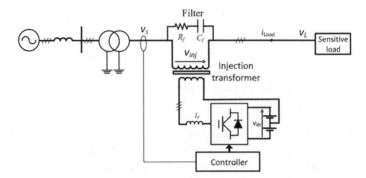

Figure 10.24   Application of a DVR to sensitive load

10.5   Series compensation

Sensitive loads such as manufacturing plants of high value products (for example, semiconductor plants or paper mills) require the voltage to be within specified limits as shown in the ITI curve in Figure 6.15. Voltage that deviates from these limits may disrupt the continuous manufacturing process, thus causing a considerable financial loss. A series-connected converter may be used, as a so-called Dynamic Voltage Restorer (DVR), to maintain the voltage within specified limits.

The connection of a DVR to a distribution feeder is shown in Figure 10.24. If the incoming feeder voltage fluctuates beyond the voltages that a sensitive load could operate, then the DVR adds a voltage in series to compensate for voltage fluctuation.

Two commonly used compensation techniques are 'in-phase compensation' and 'freeze PLL compensation'.

As shown in Figure 10.25a, the 'in-phase compensation' technique keeps the load voltage phasor always in-phase with the supply voltage. The 'in-phase compensation' technique introduces a sudden phase shift to the customer supply when the voltage sag causes a phase jump to the supply voltage vector.

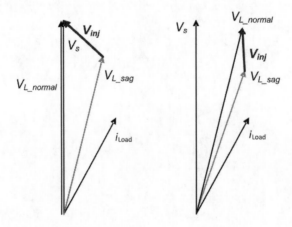

(a) In-phase compensation   (b) Freeze PLL compensation

Figure 10.25   Different compensation techniques

In the 'freeze PLL compensation' technique the DVR maintains the load voltage magnitude as the same as the pre-sag condition by injecting the difference between the pre-sag supply voltage and the sagged supply voltage (see Figure 10.25b).

Example 10.7

Part of a distribution network is shown in Figure 10.26. The load connected to busbar E is a sensitive load and therefore the voltage magnitude at that bus should be maintained at 1 ± 0.1 pu. Calculate the voltage at the load bus when a fault occurs on bus F. Using a phasor diagram, calculate the magnitude of the DVR voltage that should be injected to maintain the voltage at the sensitive load within its limits. Assume before the fault, voltage at busbar E is $1\angle 0°$ pu.

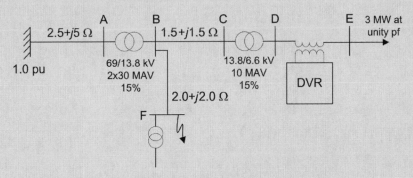

Figure 10.26   Figure for Example 10.7

Answer

All the quantities were converted to pu values based on a S_{base} of 250 MVA. Then Z_{base} at 69 kV is 19.04 Ω and Z_{base} at 13.8 kV is 0.76 Ω.

pu value of 69 kV line $= (2.5 + j5)/19.04 = 0.131 + j0.263$ pu

pu value of line BC $= (1.5 + j1.5)/0.76 = 1.97 + j1.97$ pu

pu value of line BF $= (2.0 + j2.0)/0.76 = 2.63 + j2.63$ pu

pu value of $2 \times 69/13.8$ kV transformers $= \dfrac{1}{2} \times 0.15 \times \dfrac{250}{30} = 0.625$ pu

pu value of $13.8/6.6$ kV transformer $= 0.15 \times \dfrac{250}{10} = 3.75$ pu

Load power in pu $= 3/250 = 0.012$ pu

Load could be represented by a resistance of $1/0.012 = 83.3$ pu.

The one-line representation of the distribution network shown in Figure 10.26 was redrawn using its circuit representation as shown in Figure 10.27 (this makes calculations clearer). For this calculation, the DVR was disconnected from the circuit.

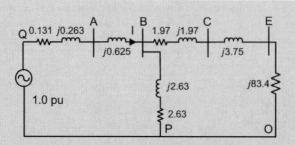

Figure 10.27   Equivalent circuit of Figure 10.26

From Figure 10.27:

$$Z_{BO} = 85.37 + j5.72 \text{pu}$$

$$Z_{QB} = 0.131 + j0.888 \text{ pu}$$

$$Z_{BO}/Z_{BP} = \cfrac{1}{\left[\cfrac{1}{85.37+j5.72} + \cfrac{1}{2.63+j2.63}\right]}$$

$$= 2.62 + j2.47 \text{ pu}$$

$$I = \left[\frac{1}{(0.131 + 2.63) + j(0.888 + 2.47)}\right] = 0.146 - j0.178 \text{ pu}$$

$$V_B = 0.823 - j0.106 \text{ pu}$$

$$V_E = (0.823 - j0.106) \times \frac{83.4}{85.37 + j5.72}$$

$$= 0.79 - j0.16 = 0.806\angle - 11.4° \text{ pu}$$

From the phasor diagram in Figure 10.28:

$$V_{inj} = \sqrt{(1 - 0.806\cos 11.4°)^2 + (0.806\sin 11.4°)^2}$$

$$= 0.263 \text{ pu} = 1.73 \text{kV}$$

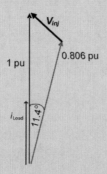

Figure 10.28   Phasor diagram

References

[1] Jenkins, N., Ekanayake, J.B. and Strbac, G. (2010) *Distributed Generation*, Institution of Engineering and Technology, Stevenage.

[2] Mohan, N., Undeland, T.M. and Robbins, W.P. (1995) *Power Electronics: Converters, Applications and Design*, John Wiley & Sons, Inc., New York.

[3] Kjaer, S.B., Pedersen, J.K. and Blaabjerg, F. (2005) A review of single-phase grid-connected inverters for photovoltaic modules. *IEEE Transactions on Industry Applications*, **41**(5), 1292–1306.

[4] Hussein, K.H., Muta, I., Hoshino, T. and Osakada, M. (1995) Maximum photovoltaic power tracking: an algorithm for rapidly changing atmospheric conditions. *IEE Proceedings Generation, Transmission and Distribution*, **142**, 59–64.

[5] King, J. and Tryfonas, J. (2009) Tidal stream power technology: state of the art, *Proceedings of Oceans Europe, IEEE Conference, Bremen, May 11–14*, pp. 1–8.

[6] Ramtharan, G., Ekanayake, J.B. and Jenkins, N. (2007) Frequency support from doubly fed induction generator wind turbines. *IET Renewable Power Generation*, **1**(1), 3–9.

[7] Lalor, G., Mullane, A. and O'Malley, M. (2005) Frequency control and wind turbine technologies. *IEEE Transactions on Power Systems*, **20**(4), 1905–1913.

[8] Ledesma, P. and Usaola, J. (2002) Contribution of variable-speed wind turbines to voltage control. *Wind Engineering*, **26**(6), 347–358.

[9] Fraile-Ardanuy, J., Wilhelmi, J.R., Fraile-Mora, J.J. and Perez, J.I. (2006) Variable-speed hydro generation: operation and contol. *IEEE Transaction Energy Conversion*, **21**(2), 569–574.

[10] *Status Report on Variable Speed Operation in Small Hydropower*, EU project ENERGIE, 2000, http://ec.europa.eu/energy/res/sectors/doc/small_hydro/statusreport_vspinshp_colour2.pdf (accessed on 4 August 2011).

[11] Hamidi, V., Smith, K.S. and Wilson, R.C. (2010) *Smart Grid Technology Review within the Transmission and Distribution Sector*, Innovative Smart Grid Technologies Europe, Sweden.

[12] Renz, K., Thumm, G. and Weiss, S. (1995) Thyristor control for fault current limitation. IEE Colloquium on Fault Current Limiters – A Look at Tomorrow, 1995, pp. 3/1–3/4.

[13] Putrus, G.A., Jenkins, N. and Cooper, C.B. (1995) A static fault current limiting and interrupting device. IEE Colloquium on Fault Current Limiters – A Look at Tomorrow, 1995, pp. 5/1–5/6.

[14] Ahmed, M.M.R. (2008) Comparison of the performance of two solid state fault current limiters in the distribution network. 4th IET Conference on Power Electronics, Machines and Drives, 2008, pp. 772–776.

[15] *IEEE Standard 519-1992: IEEE Recommended Practices and Requirements for Harmonic Control in Electrical Power Systems*, IEEE, 1992.

[16] *Harmonic Standard ER G5/4*, Engineering Recommendation, 2001.

[17] Ekanayake, J.B. and Jenkins, N. (1996) A three-level advanced static VAr compensator. *IEEE Transactions on Power Delivery*, **11**(1), 540–545.

[18] Schauder, C. and Mehta, H. (1993) Vector analysis and control of Advanced Static VAr compensators. *IEE Proceedings-C*, **140**(4), 299–306.

[19] Akagi, H., Kanazawa, Y. and Nabae, A. (1984) Instantaneous reactive power compensators comprising switching devices without energy storage components. *IEEE Transactions on Industry Applications*, **1A-20**(3), 625–630.

[20] Ghosh, A. and Ledwich, G. *Applications of Power Electronics to Power Distribution Systems*, Document No 05TP176, IEEE.

[21] Hingorani, N.G. and Gyugyi, L. (1999) *Understanding FACTS: Concepts and Technology of Flexible AC Transmission Systems*, IEEE Power Engineering Society, New York.

[22] Saad Saoud, Z., Lisboa, M.L., Ekanayake, J.B. *et al.* (1998) Application of STATCOMs to wind farms. *IEE Proc. Generation, Transmission and Distribution*, **145**(5), 511–516.

[23] Schoenung, S.M. and Burns, C. (1996) Utility energy storage application studies. *IEEE Transactions on Energy Conversion*, **11**(3), 658–665.

[24] Baker, J.N. and Collinson, A. (1999) Electrical energy storage at the turn of the millennium. *Power Engineering Journal*, **13**(3), 107–112.

11

Power Electronics for Bulk Power Flows

11.1   Introduction

The future power system will involve connection of a great number of large renewable energy schemes and other new low-carbon generators that will be needed to reduce emissions and maintain the continued security of supply. These new connections and subsequent bulk power flows will require network reinforcement. The traditional methods of increasing bulk power transfer capacity are reconducting existing circuits, upgrading to a higher AC voltage and constructing new lines. However, these options, particularly ones which involve new overhead line routes, are difficult to implement due to planning constraints and environmental concerns. FACTS devices can increase the capacity of AC circuits while HVDC, especially submarine routes, may be used for the addition of new capacity.

Figure 11.1 (also refer to Plate 7) [1] shows an example of network reinforcement using FACTS and HVDC. This proposed network reinforcement in northern Britain to bring wind-generated electricity to London includes two submarine cable circuits and extensive series capacitor compensation on the terrestrial 400 kV AC overhead lines. The East Coast HVDC link is being constructed as a CSC-HVDC while the West Coast HVDC link may use multi-terminal VSC-HVDC.

More speculatively, an offshore HVDC grid has been proposed that builds on the 11 submarine circuits currently operating and 21 others being considered across the North and Baltic Seas. This will lead to a European-wide HVDC network interconnecting about 300 GW of wind generation on the north and west coasts of Europe and could extend to include 700 GW of solar generation around the Mediterranean [2]. It is anticipated that with some 200 GW of hydro in the Nordic regions, the proposed interconnected network could ensure regional smoothing of electrical energy generation, increased security of supply and reduced dependency on fuel imports. Balancing of supply and demand would benefit from the diversity of the renewable energy resources and of the load through the spread of time zones.

Smart Grid: Technology and Applications, First Edition.
Janaka Ekanayake, Kithsiri Liyanage, Jianzhong Wu, Akihiko Yokoyama and Nick Jenkins.
© 2012 John Wiley & Sons, Ltd. Published 2012 by John Wiley & Sons, Ltd.

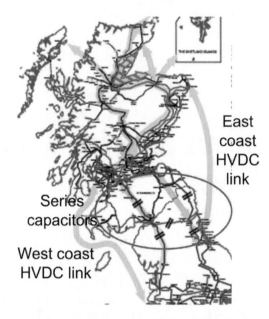

Figure 11.1 Proposed reinforcements to the UK Transmission network [1]

11.2 FACTS

FACTS devices can enhance the power flow on existing power lines. For the transmission line shown in Figure 11.2, the sending end voltage is $V_S \angle \delta_S$, the receiving end voltage is $V_R \angle \delta_R$ and the equivalent impedance of parallel connected lines is X. The power transfer through the lines is given by:

$$P = \frac{V_S V_R}{X} \sin (\delta_S - \delta_R) \tag{11.1}$$

Figure 11.2 also shows how FACTS devices act on the power transfer equation. The TCSC can change the impedance of the line, the STATCOM can control the voltage magnitude at

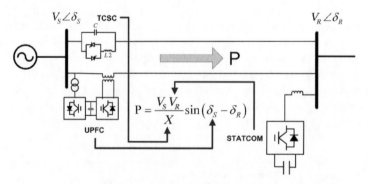

Figure 11.2 FACTS applications for increased power transfer [3]

the terminal to which it is connected by injecting or absorbing reactive power and the UPFC can alter the phase angle of the sending end voltage, thus power flow through a line can be controlled in a number of ways.

11.2.1   Reactive power compensation

In transmission circuits as the load at the receiving end varies, the receiving end voltage also changes. To maintain the voltage within limits, reactive power compensation devices may be employed. They control the voltage at a node by injecting or absorbing reactive power. In addition to steady state voltage control, reactive power compensation devices are used to mitigate dynamic voltage variations caused by line switching, load rejection, faults and other disturbances. As indicated in Figure 11.2, they can also increase the power transfer capability of a line and improve stability [4].

Reactive power compensation is now becoming essential to meet the requirements set by many utilities for the connection of renewable energy generators. In many countries the national Grid Code, which defines the minimum operational and technical requirements for connection of generation plant to the grid, stipulates steady state and dynamic reactive power requirements. These requirements need to be satisfied at the point of connection of the generating plant to the utility network. With a wind farm, reactive power compensation is often required at the point of connection to satisfy the Grid Code reactive power requirements. Reactive power compensation devices may also be needed to meet fault ride through requirements [5].

Figure 11.3 shows the shunt reactive power compensation devices commonly found in the power system. Fixed capacitors provide a constant capacitive shunt reactance to the network (in some Applications, a shunt inductor is used to obtain an inductive reactance). A reactor can be connected in series with the fixed capacitor to form a harmonic filter (Figure 11.3a). The filter can be designed to draw the same fundamental current as the fixed capacitor at the system frequency and provide low-impedance shunt path at the harmonic frequencies.

The Thyristor Controlled Reactor (TCR) and Thyristor Switched Capacitor (TSC) give a variable reactive impedance. The reactive power output of these is proportional to the square of the line voltage divided by the reactive impedance. A STATCOM provides a variable reactive current that is injected into the transmission network, thus its reactive power output is the product of the line voltage and reactive current injected/absorbed.

The basic elements of a TCR are a reactor in series with a bi-directional thyristor pair as shown in Figure 11.3b [4, 6]. Each thyristor conducts on alternative half-cycles of the supply

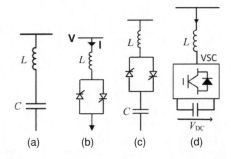

(a)   (b)   (c)   (d)

Figure 11.3   Commonly used reactive power compensators

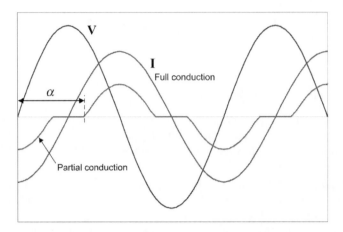

Figure 11.4 Voltage across and current through TCR

frequency. The current flow in the inductor (L) is controlled by adjusting the conduction interval of the back to back-connected thyristors. This is achieved by delaying the closure of the thyristor switch by an angle α, the firing angle, in each half cycle with respect to the voltage zero. As shown in Figure 11.4, when $\alpha = 90°$, the current is essentially reactive and sinusoidal. Partial conduction is obtained by firing angles between 90 and 180°.

With partial conduction, the TCR generates harmonics. If the gating angles are balanced, only odd harmonics are generated. In three-phase applications the basic TCR elements are connected in delta, thus no triplen harmonic are present in the line currents. The lower order harmonics up to the 9th can be eliminated by using two delta connected TCRs of equal rating fed from two secondary windings of the step-down transformer, one connected in star and other in delta. This forms a 12-pulse TCR as shown in Figure 11.5.

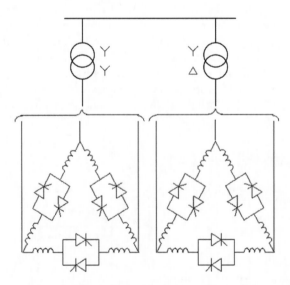

Figure 11.5 A 12-pulse TCR

Figure 11.6   A TSC–TCR arrangement. *Source:* Photo courtesy of Toshiba

The basic elements of a TSC are a capacitor in series with a bi-directional thyristor pair and a small reactor as shown in Figure 11.3c [4,6]. The purpose of the reactor is to limit switching transients, to damp inrush currents and to be a filter for harmonics coming from the power system. The capacitance is adjusted by controlling the number of parallel capacitors connected in shunt. Each capacitor always conducts for an integral number of half cycles.

The switching of capacitors excites transients and it is necessary to switch the capacitors at a point where the switching transients are minimum [4,6]. This is achieved by pre-charging the capacitor to the crest value of the supply voltage and by choosing the switching instant when the voltage across the thyristor switch is minimum, that is, at the crest value of the supply voltage. The switch-off period corresponds to a current zero after an integral number of half cycles.

A TCR only absorbs reactive power, so in many applications a fixed capacitor or a TSC is connected in parallel with the TCR. A TSC–TCR arrangement is shown in Figure 11.6 (also refer to Plate 6). This combined arrangement is commonly referred to as a Static Var Compensator (SVC).

As discussed in Section 10.4.1, the STATCOM shown in Figure 11.3d is a voltage source converter whose DC side is connected to a capacitor and whose AC side is connected to the grid via a transformer having high reactance or a transformer with a series connected reactor. A STATCOM uses IGBTs or GTOs as switching devices and can generate or absorb reactive power. Its dynamic response is faster than a TSC or TCR and allows continuous control of reactive power.

Example 11.1

A $\pm$ 30 Mvar Static Var Compensator is connected to 132 kV system via a 132/6 kV transformer. The SVC consists of a 12-pulse TCR and two switched capacitor arms, one tuned to the 11th harmonic and the other tuned to the 13th harmonic (the lowest harmonics produced by the TCR) and producing equal amounts of reactive power.

Calculate the values of the passive components required for the TCR and the switched capacitor branches.

Answer

For each delta connected TCR section of the 12-pulse TCR, $Q = 15$ MVar.
From

$$Q = \frac{\sqrt{3}V_{LL}^2}{X} : 15 \times 10^6 = \frac{\sqrt{3} \times (6 \times 10^3)^2}{X_L}$$

Therefore, $X_L = 2\pi \times 50 \times L = 4.16\ \Omega$ and hence $L = 13.2$ mH.
As the 1st capacitor branch is tuned to 11th harmonic component

$$\frac{1}{\sqrt{L_{11}C_{11}}} = 2\pi \times 50 \times 11$$

$$\therefore L_{11}C_{11} = 83.74 \times 10^{-9} \tag{11.2}$$

At 50 Hz, this filter arm should produce capacitive reactive power of 15 Mvar.
The impedance of this filter at 50 Hz =

$$\frac{1}{100\,\pi C_{11}} - 100\,\pi L_{11}$$

From

$$Q = \frac{\sqrt{3}V_{LL}^2}{X} : \frac{\sqrt{3} \times (6 \times 10^3)^2}{\left(\frac{1}{100\,\pi C_{11}} - 100\,\pi L_{11}\right)} = 15 \times 10^6$$

$$\therefore \frac{1}{100\,\pi C_{11}} - 100\,\pi L_{11} = 4.16 \tag{11.3}$$

From Equations (11.2) and (11.3):

$$L_{11} = 110\ \mu\text{H} \qquad C_{11} = 759\ \mu\text{F}$$

The 2nd capacitor branch is tuned to the 13th harmonic component

$$\frac{1}{\sqrt{L_{13}C_{13}}} = 2\pi \times 50 \times 13$$

$$\therefore L_{13}C_{13} = 59.95 \times 10^{-9} \tag{11.4}$$

At 50 Hz, this filter arm should produce capacitive reactive power of 15 Mvar.
The impedance of this filter at 50 Hz =

$$\frac{1}{100\,\pi C_{13}} - 100\,\pi L_{13}$$

From

$$Q = \frac{\sqrt{3}V_{LL}^2}{X} : \frac{\sqrt{3} \times (6 \times 10^3)^2}{\left(\frac{1}{100\pi C_{13}} - 100\pi L_{13}\right)} = 15 \times 10^6$$

$$\therefore \frac{1}{100\pi C_{13}} - 100\pi L_{13} = 4.16 \qquad\qquad (11.5)$$

From Equations (11.4) and (11.5):

$$L_{13} = 78.7 \ \mu H \qquad C_{13} = 761 \ \mu F$$

Example 11.2

Ten 2 MW wind turbines are connected to the network shown in Figure 11.7a. Each wind turbine has the capability chart of Figure 11.7b measured at 33 kV. All the wind turbines in the wind farm are operating at point A of Figure 11.7b. Calculate the reactive power that should be absorbed or generated by a STATCOM connected at the point of connection to maintain the power factor at the point of connection at 0.95 exporting vars. Ignore the active and reactive power losses within the wind farm and assume that the voltage at the wind farm busbar (W) is $1.05\angle 0°$ pu.

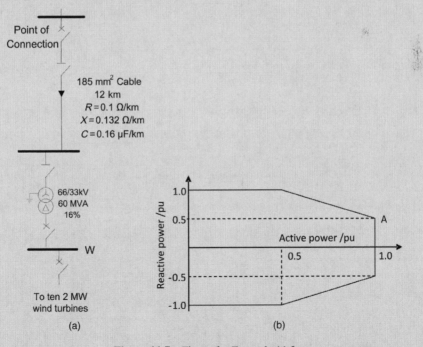

Figure 11.7   Figure for Example 11.2

Answer

On $S_{base} = 100\,\text{MVA}$:

Active power generated by the wind farm $= 10 \times 2/100 = 0.2$ pu.

Reactive power generated by the wind farm $= 10 \times 0.5 \times 2/100 = 0.1$ pu

$$Z_{base}\text{on 100 MVA, 66 kV} = \frac{\left(66 \times 10^3\right)^2}{100 \times 10^6} = 43.56\,\Omega$$

Line impedance

$$\frac{(0.1 + j0.132) \times 12}{43.56} = 0.028 + j0.036\,\text{pu}$$

Transformer reactance on 100 MVA base $= 0.16 \times 100/60 = 0.267$ pu. The equivalent circuit can be drawn as shown in Figure 11.8:

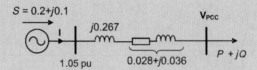

Figure 11.8   Equivalent circuit of Figure 11.7

$$\mathbf{VI}^* = 0.2 + j0.1$$

$$\mathbf{I} = \frac{0.2 - j0.1}{1.05}$$

$$= 0.19 - j0.095$$

$$\mathbf{V_{PCC}} = 1.05 - (0.19 - j0.095) \times (j0.267 + 0.028 + j0.036)$$

$$= 1.05 - (0.19 - j0.095)(0.028 + j0.303)$$

$$= 1.016 - j0.055$$

$$= 1.017\angle - 3.1°$$

Therefore,

$$P + jQ = \mathbf{V_{PCC}I}^* = 1.017\angle - 3.1° \times (0.19 + j0.095)$$

$$= 1.017\angle - 3.1° \times 0.212\angle26.6°$$

$$= 0.216\angle23.5°$$

$$= 0.198 + j0.086$$

Therefore the power factor angle $= \tan^{-1}(0.086/0.198) = 23.5°$

The power factor angle for a power factor of $0.95 = \cos^{-1}0.95 = 18.2°$

The reactive power that should be absorbed by the STATCOM to get a power factor of 0.95 exporting vars at the point of connection =

$$0.198 \times \left[\tan 23.5° - \tan 18.2°\right]$$

$$= 0.021 \text{ pu} = 2.1 \text{ Mvar}$$

11.2.2   Series compensation

Connection of capacitors or inductors in shunt changes the flow of reactive power in a circuit and so changes the network voltage. In general, a modest level of shunt reactive compensation is benign and poses few risks to the power system. Similarly, power electronic shunt compensators alter the reactive power flows. In contrast, capacitors may be connected to a circuit in series to reduce its inductive reactance. This can be a more hazardous approach as very high voltages can occur across the capacitor during faults and electro-mechanical resonances can be stimulated in the shafts of rotating machines [7]. For a given value of capacitance, series compensation altering the reactance of the circuit is more effective in controlling voltages than shunt compensation that changes the reactive power flows.

The simplest form of series reactive power compensation device is a series-connected capacitor. Figure 11.9 shows the thyristor-based series reactive power compensation devices found in the power system. These series compensators rely on their control system to avoid resonances and ensure the stability of the power system.

In a Thyristor Switched Series Capacitor (TSSC) (Figure 11.9a), the series capacitance is varied by switching on and off the segments of capacitors in series with the line. The thyristors are fully conducting or fully blocked for a number of half-cycles as determined by the control system. The TSSC only permits the insertion of capacitance to be controlled in discrete steps.

The Thyristor Controlled Series Capacitor (TCSC) is a series capacitor in parallel with a thyristor controlled reactor as shown in Figure 11.9b. The effect of this parallel scheme is to form a continuous and rapidly variable series impedance. Figure 11.10 shows the change in impedance with the firing angle delay of the thyristors. When $\alpha = 90°$ (alpha defined as in Figure 11.4), the thyristors are fully conducting and when $\alpha = 180°$, the thyristors are fully blocked. As shown in Figure 11.4, partial conduction of the thyristor arm is obtained with firing angles between 90 and 180°. Therefore the thyristor arm effectively provides a variable inductor in parallel with the capacitor. When the inductive reactance is equal to the capacitive

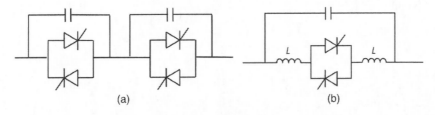

(a)   (b)

Figure 11.9   Thyristor-based series compensation

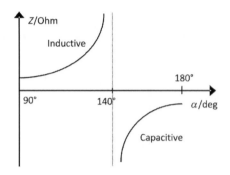

Figure 11.10 TCSC characteristic

reactance, the TCSC goes into parallel resonance as shown in Figure 11.10. At this point (for this TCSC at α around 140°), the impedance changes rapidly from inductive to capacitive.

Example 11.3

A series compensated transmission line is shown in Figure 11.11. On 100 MVA, 400 kV base $X_g = 0.1$ pu, $X_T = 0.02$ pu, $R_L = 0.0015$ pu, and $X_L = 0.04$ pu. Calculate the value of the series capacitor and electrical resonance frequency for 30 per cent and 60 per cent compensation. Generator resistance can be ignored.

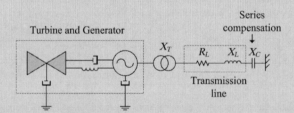

Figure 11.11 Figure for Example 11.3

Data: percentage compensation is defined as: $K = (X_C)/(X_g + X_T + X_L) \times 100$ [7].

Answer

At 30 per cent compensation:

$$\frac{X_C}{X_g + X_T + X_L} \times 100 = 30$$

$$\frac{X_C}{0.1 + 0.02 + 0.04} = 0.3$$

$$X_C = 0.048 \text{ pu}$$

Base impedance =

$$= \frac{\left(400 \times 10^3\right)^2}{100 \times 10^6} = 1600 \ \Omega$$

$$\therefore X_C = \frac{1}{2\pi fC} = 0.048 \times 1600 = 76.8 \ \Omega$$

With $f = 50$ Hz, C = 41.4 μF.
Series resonance frequency, f_C, is given by:

$$f_C = \frac{1}{2\pi}\sqrt{\frac{1}{LC}} = \frac{f}{2\pi f}\sqrt{\frac{1}{LC}} = f\sqrt{\frac{1}{\omega^2 LC}} = f\sqrt{\frac{X_C}{X_d'' + X_T + X_L}} = \frac{f\sqrt{K}}{10}$$

Therefore, with 30 per cent compensation $f_C = (50\sqrt{30})/10 = 27.4$ Hz
At 60 per cent compensation:

$$\frac{X_C}{X_d'' + X_T + X_L} \times 100 = 60$$

$$\frac{X_C}{0.1 + 0.02 + 0.04} = 0.6$$

$$X_C = 0.096 \text{ pu}$$

$$\therefore X_C = \frac{1}{2\pi fC} = 0.096 \times 1600 = 153.6 \ \Omega$$

With $f = 50$ Hz, $C = 20.7$ μF and $f_C = (50\sqrt{60})/10 = 38.7$ Hz
If the mechanical natural frequency of oscillation of the generator and turbine on its shaft is close to one of these electrical resonant frequencies, then severe damage can result.

11.2.3   Thyristor-controlled phase shifting transformer

Phase shifters are widely used in power systems for controlling the magnitude and direction of the active power flow, often over parallel circuits. The principle of operation of the conventional phase shifter is explained in Figure 11.12a. Control of the magnitude and direction of active power flow on the line is achieved by injecting a voltage in series with the line, thus changing the phase angle of the receiving end voltage. A variable series voltage is obtained by a tap-changer acting on the regulating winding. This voltage is in quadrature with the input voltage. It is then injected by the booster winding across the series winding [8].

Rapid phase angle control can be accomplished by using a thyristor switching network to vary the injected voltage. One possible arrangement is shown Figure 11.12b. In this case one

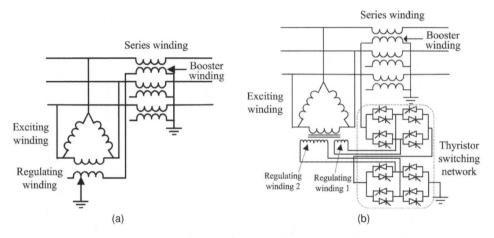

Figure 11.12   Phase shifter. *Note:* Only single phase connection to the booster winding is shown

choice would be to select the number of turns in two regulating windings in the ratio 1 : 3. When winding 1 is connected, a series voltage, say, V, is injected in series with the line. The connection could be reversed, thus getting $-V$ injected voltage. Winding 2 (gives injected voltage of $\pm 3\ V$) and winding 1 (gives injected voltage of $\pm V$) could be used to generate injected voltages from $-4\ V$ to $+4\ V$ in steps of V.

Example 11.4

A 400 kV transmission line has a reactance of 0.05 pu on 100 MVA base. The voltage at the sending end is $1.02\angle 2°$ and that of the receiving end is $1.0\angle 0°$. What is the voltage that should be injected in quadrature with the sending end voltage to increase the power transfer across the line by 20 per cent?

Answer

The power transfer through the lines is:

$$P = \frac{V_S V_R}{X} \sin(\delta_S - \delta_R)$$

$$= \frac{1.02 \times 1}{0.05} \sin(2° - 0)$$

$$= 0.71\ \text{pu}$$

If the power transfer is increased by 20 per cent, P = 0.85 pu. Assuming that the sending voltage phasor is constant, from new power transfer:

$$0.85 = \frac{1.02 \times 1}{0.05} \sin(2° - \delta_R)$$

$$\therefore \delta_R = -0.4°$$

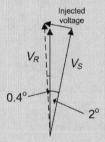

Figure 11.13 Phasor diagram

From the phasor diagram (Figure 11.13):
Injected voltage =

$$V_S \tan(2.4°) = 1.02 \tan(2.4°) = 0.04 \, \text{pu}$$

11.2.4 Unified power flow controller

The Unified Power Flow Controller (UPFC) is an AC to AC converter which can change the parameters of a circuit: line end voltage, the phase angle between the two busbars and the apparent reactance of the line. It is implemented by two AC to DC converters operating from a common DC link capacitor as shown in Figure 11.14. One converter is connected in series and the other is in shunt with the transmission line. VSC2 generates a voltage $V_x(t) = V \sin(\omega t - \alpha)$ at the fundamental frequency (ω) with variable amplitude ($0 \leq V \leq V_{\text{max}}$) and

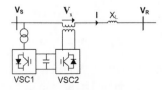

Figure 11.14 UPFC

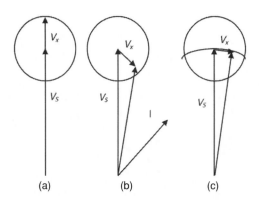

Figure 11.15   UPFC operation

phase angle ($0 \le \alpha \le 2\pi$) which is added to the AC system voltage by the series connected coupling transformer.

From Figure 11.14, the receiving end apparent power is given by:

$$\mathbf{S_R} = \mathbf{V_R I^*} = \mathbf{V_R} \left[\frac{\mathbf{V_S} + \mathbf{V_x} - \mathbf{V_R}}{jX} \right]^*$$

$$= \mathbf{V_R} \left[\frac{\mathbf{V_S} - \mathbf{V_R}}{jX} \right]^* + \mathbf{V_R} \left[\frac{\mathbf{V_x}}{jX} \right]^* \qquad (11.6)$$

The first term in Equation (11.6) is the apparent power associated with the uncompensated line and the second term is the apparent power associated with the series injection by VSC2. The real power demanded by VSC2 is supplied by VSC1 through the DC link capacitor. VSC1 can also act as a shunt reactive power compensator, thus injecting or absorbing reactive power independent of the operation of the series connected converter, VSC2.

As shown in Figure 11.15, VSC2 can be controlled to obtain terminal voltage regulation, line impedance control or phase angle control. Terminal voltage regulation is achieved by injecting a voltage in phase with the sending end voltage as shown in Figure 11.15a. The line impedance can effectively be varied by injecting a voltage perpendicular to the line current (see Figure 11.15b). Phase angle control can be achieved by injecting a voltage to shift the sending end voltage by an angle (see Figure 11.15c).

11.2.5   Interline power flow controller

An Interline Power Flow Controller (IPFC) connects a number of VSCs in series to different lines. The DC side is connected in parallel as shown in Figure 11.16. This enables power to be transferred from one line to another through the series links.

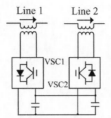

Figure 11.16   IPFC connected to two lines

Example 11.5

A 500 MW wind farm is connected to two transmission regions as shown in Figure 11.17. At a given instant the wind farm supplies 200 MW to each region. When the wind farm output increases to 450 MW, it is necessary to send 250 MW to region 1 and 200 MW to region 2. Discuss how an IPFC could be used to achieve this. Assume the voltage magnitude and angle of the busbars in regions 1 and 2, to which the tie line is connected, are unchanged.

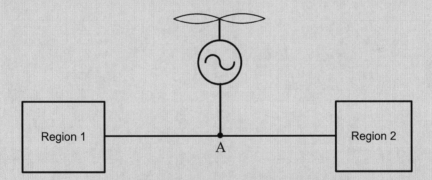

Figure 11.17   Figure for Example 11.5

Answer

When the wind farm output power increases to 450 MW, if the line is uncompensated, then 225 MW will flow to each region. An IPFC could be connected to point A so that it takes 25 MW from Line 2 and feed to Line 1 as shown in Figure 11.18a. The way the IPFC controller could achieve this is described by the phasor diagram shown in Figure 11.18b. For clarity, it was assumed that Lines 1 and 2 have equal reactance, X. VSC1 injects a voltage in series with the line so as to increase the load angle from δ to δ_1, thus increasing the power flow in Line 1. On the other hand, VSC2 injects a voltage in series with the line so as to reduce the load angle from δ to δ_2, thus reducing the power flow in Line 2.

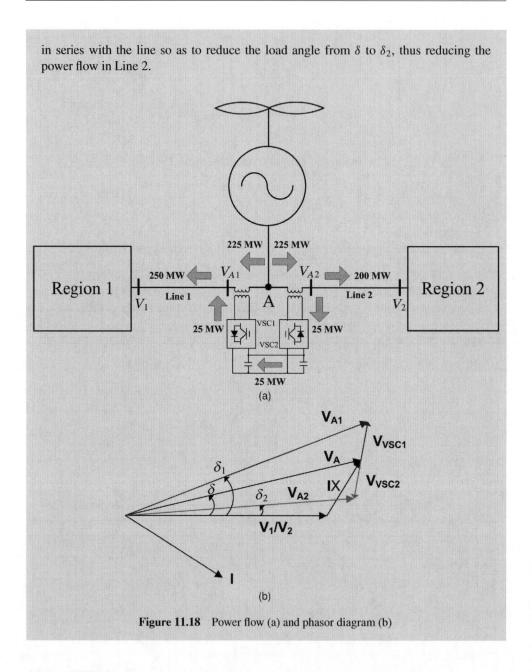

Figure 11.18   Power flow (a) and phasor diagram (b)

11.3   HVDC

Transmission of power using HVDC to connect national electricity grids, or large generation plants to load centres has existed for the past 50 years. For large cities with high load densities that depend on power imports from outside, HVDC links are an ideal choice due to the security and flexibility they provided. For example, the 400 MW VSC HVDC link between Pittsburg

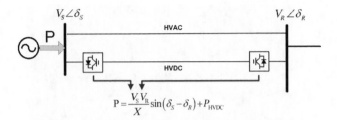

Figure 11.19 Influence of HVDC on power transfer

and San Francisco was constructed to minimise the transmission capacity constraints in San Francisco.

HVDC systems are also connected in parallel with AC circuits to enhance the flexibility of AC transmission lines. The HVDC Pacific Intertie, which is connected in parallel with the AC intertie between the Pacific Northwest to the Los Angeles area of the USA, serves as a flexible link to transmit power from north to south of the USA.

Figure 11.19 shows a VSC-based HVDC connected in parallel with an AC line. In addition to providing real power flow from the sending to the receiving end, the two converters can also control their reactive power so as to control the voltage magnitude at the sending or receiving end thus influencing the AC power flow.

For the connection of large renewable energy sources, such as offshore wind farms, which are normally far away from load centres (typically the best sites for renewable energy generation are located a long distance from load centres), HVDC can be attractive as it is flexible and economic.

11.3.1   Current source converters

The Current Source Converter HVDC (CSC-HVDC) has been the technology of choice to transmit large amounts of power from one point to another. The basic building block of both rectifier and the inverter of the CSC-HVDC is a current source converters [9]. CSC-HVDC technology is primarily chosen because of the reliability and robustness inherent within the thyristor valves at the heart of the converter. A thyristor valve is shown in Figure 11.20 (also refer to Plate 8).

The nature of the thyristor's switching operation leads to a fundamental limitation in conventional CSC-HVDC equipment. The converters always draw reactive power, require a strong AC voltage source for each converter and are susceptible to inverter commutation failure. Over the past 50 years, there have been a number of excellent books written on CSC-HVDC [9, 10, 11] and the subject is dealt with briefly here. A schematic of a CSC-HVDC scheme and its equivalent circuit are shown in Figure 11.21. The equivalent circuit was obtained using the rectifier equivalent circuit shown in Figure 9.7 and a similar equivalent circuit for the inverter. For the inverter as the firing angle, α, is greater than $\pi/2$, a new angle, $\beta = \pi - \alpha$ is defined.

From the equivalent circuit shown in Figure 11.21b:

$$I_d = \frac{1.35V_{LL}\cos\alpha - 1.35V_{LL}\cos\beta}{\frac{3\omega L_r}{\pi} + \frac{3\omega L_i}{\pi} + r_L} \tag{11.7}$$

Figure 11.20   Thyristor valve. *Source:* Courtesy of Toshiba

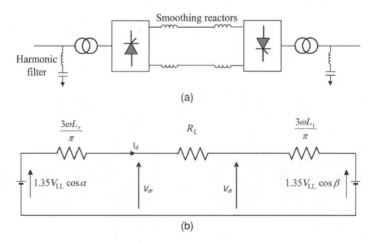

Figure 11.21   A CSC-HVDC scheme

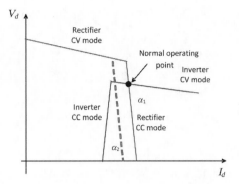

Figure 11.22 CSC-HVDC control

The typical control approach used for steady state power transfer would be that the rectifier maintains a Constant Current (CC) that is, its controller changes α with the load. The inverter maintains a Constant Voltage (CV) that is, β is kept constant. As the denominator of Equation (11.7) is very small, a small change in α results in a large change in I_d. As I_d increases with more power transfer through the link, α should decrease proportionally and as α reaches its minimum limit (about $5°$), no further control is possible and the rectifier is operated at constant α, that is CV mode. This control philosophy is shown in Figure 11.22.

Example 11.6

A 12-pulse rectifier is shown in Figure 11.23. The effective turns ratio, n, of the transformer is 0.4. When the primary voltage is 220 kV, the firing angle delay is $15°$, the DC current delivered by the rectifier is 1000 A. Calculate (1) DC link voltage; (2) rms current; and (3) reactive power absorbed by the 12-pulse converter.

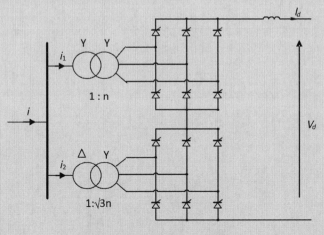

Figure 11.23 A 12-pulse rectifier

Answer

1. For a single six-pulse converter:

$$\frac{V_d}{2} = 1.35 V_{LL} \cos \alpha$$

$$= 1.35 \times 220 \times 0.4 \times \cos 15°$$

$$= 114.75 \text{ kV}$$

Therefore, the DC link voltage $= 114.75 \times 2 = 229.5$ kV.

Note that even though one converter is connected through a star-delta transformer, due to the turns ratio of $1:\sqrt{3}n$ of that converter transformer, on the DC side both converters act identically.

2. The rms value of current i_1

$$= \frac{\sqrt{6}}{\pi} \times n \times I_d = \frac{\sqrt{6}}{\pi} \times 0.4 \times 1000 = 311.88 \, A.$$

Even though i_2 is through a delta-star transformer the way the transformer ratio is selected, the rms value of i_2 also is equal to 311.88 A. Therefore the rms value of current $i = 2 \times 311.88 \text{ A} = 623.76 \text{ A}$.

3. As the power factor angle is approximately equal to the firing angle $\phi = 15°$.

Reactive power absorbed by the 12-pulse converter

$$= \sqrt{3} V I \cos \phi = \sqrt{3} \times 220 \times 10^3 \times 623.76 \times \sin 15°$$
$$= 61.5 \, MVar$$

Example 11.7

A HVDC link delivers 1000 MW. Its DC voltage is 500 kV. The DC line resistance is 2 Ω/line and each 6-pulse converter has an equivalent commutating resistance of 2 Ω. The DC link is operating with the rectifier on CC mode with $\alpha = 15°$ and the inverter on CV mode with $\beta = 20°$. Calculate the following:

1. RMS value of the line-to-line voltage.
2. Reactive power absorbed by the inverter.

Answer

1. DC current $= 1000 \times 10^6 / 500 \times 10^3 = 2000$ A
   From Equation (11.7):

$$I_d = 2000 = \frac{1.35 V_{LL} \cos 15° - 1.35 V_{LL} \cos 20°}{2 + 2 + 2 \times 2}$$

$$V_{LL} = 451.8 \text{ kV}$$

2. Since the power factor angle is equal to the firing angle, $\phi = 15°$.

The RMS current $= \dfrac{\sqrt{6}}{\pi} I_d = \dfrac{\sqrt{6}}{\pi} \times 2000 = 1559.4\,A$

Reactive power absorbed

$$= \sqrt{3}VI \sin\phi = \sqrt{3} \times 451.8 \times 10^3 \times 1559.4 \times \sin 15°$$
$$= 315.8 \text{ Mvar}$$

11.3.2   Voltage source converters

Increases in the voltage and current ratings of self-commutating power electronic devices now allow the use of a Voltage Source Converters (VSCs) for HVDC. The full controllability through both turn-on and turn-off operation of the IGBTs allows the self-commutated VSCs to reverse power flow much more quickly than CSC and eliminates the risk of commutation failure. The VSC also has the ability to absorb and generate both active and reactive power independently of one another. Further advantages are that the generation of harmonics is greatly reduced, minimising the footprint of filters required to absorb them and the capability to blackstart an AC system (that is, restore power without the aid of an external voltage source).

Typical components of a VSC are shown in Figure 11.24. These include:

1. *Valve Units*: The main component of these valve units is the switching device that is utilised to control the converter, the most common of these devices being the IGBT. To increase the voltage rating of the valve unit, a large number of IGBTs are connected in series. Normally a number of redundant devices are built into the valve to increase its robustness and delay any maintenance until the next routine maintenance period.

   For the converter to work effectively auxiliary circuitry is required for each IGBT. Snubbers to minimise over-voltages, an over-current protection circuit, and voltage-sharing resistors are usually connected in parallel to the IGBT. Further, the gate of each switch needs control circuitry.

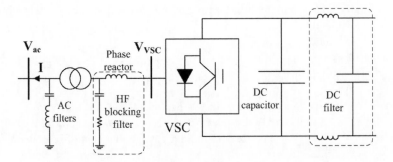

Figure 11.24   Significant components of a VSC-SCC

2. *DC Capacitors*: The primary function of the capacitors on the DC side of the converter is to stabilise the DC voltage. Sizing these capacitors should be in accordance with the switching frequency for optimum performance and economy, the faster the switching frequency, the smaller the capacitors can be.
3. *AC Transformers*: The transformer provides isolation and voltage matching. It also contributes to the reactance between the VSC and the grid.
4. *Phase Reactors*: The phase reactors serve several purposes; they assist in controlling the flow of reactive and active power, reduce harmonic currents and limit any fault currents.
5. *Filters*: AC filters are used in VSC to reduce higher-order harmonics produced by the switching of the converters. As these filters are not required to provide reactive power (unlike in the case of CSC technology), their size is significantly reduced. HF blocking filters protect the transformer from high dv/dt stresses and prevent DC entering the transformers. The DC filter reduces the harmonic currents on the DC side.

From Figure 11.24, if harmonic voltages produced by the VSC are ignored, the line current can be written as:

$$\mathbf{I} = \frac{\mathbf{V_{VSC}} - \mathbf{V_{ac}}}{jX} \tag{11.8}$$

where X is the total reactance of the phase reactor and the transformer. Therefore, the power flow from VSC-HVDC to the grid is given by:

$$P + jQ = \mathbf{V_{ac}I^*} = \mathbf{V_{ac}} \left[\frac{\mathbf{V_{VSC}} - \mathbf{V_{ac}}}{jX} \right]^* \tag{11.9}$$

Assuming that the voltage produced by the VSC is $V_{VSC}\angle\delta$ (therefore $\mathbf{V_{VSC}} = V_{VSC}(\cos\delta + j\sin\delta)$) with respect to $\mathbf{V_{ac}}$, P and Q flows can be obtained from Equation (11.9) as:

$$P = \frac{V_{ac}V_{VSC}\sin\delta}{X} \tag{11.10}$$

$$Q = \frac{V_{ac}(V_{VSC}\cos\delta - V_{ac})}{X} \tag{11.11}$$

The magnitude (V_{VSC}) and phase angle (δ) of the voltage that should be produced at the terminal of the VSC to obtain reference active and reactive power flows can be obtained from Equations (11.10) and (11.11).

Example 11.8

A 500 MW VSC is connected to a 400 kV network through a 600 MVA, 400/250 kV transformer having a leakage reactance of 20 per cent. The reactance of the phase reactor on 500 MVA, 250 kV base is 0.2 pu. Find the voltage that should be generated by the

VSC to deliver 300 MW to the grid at unity power factor. If the DC link voltage is 500 kV, what should be the value of the modulation index?

Data: For the VSC operating on sinusoidal PWM $V_{LL} = 0.612 \times m_a \times V_{DC}$ where V_{LL} is the line-to-line voltage at D-STATCOM terminals, m_a is the modulation index and V_{DC} is the DC capacitor voltage.

Answer

On a 500 MVA base, the reactance of the transformer $= 0.2 \times 500/600 = 0.167$ pu.
Active power that should be delivered $= 300/500 = 0.6$ pu.
From Equation (11.10):

$$P = \frac{1 \times V_{VSC} \sin \delta}{(0.167 + 0.2)} = 0.6$$

$$V_{VSC} \sin \delta = 0.22$$

As the power is delivered at unity power factor, $Q = 0$; thus, from Equation (11.11):

$$V_{VSC} \cos \delta - V_{ac} = 0$$
$$V_{VSC} \cos \delta = 1$$

Therefore $V_{VSC} = \sqrt{0.22^2 + 1} = 1.024$ pu $= 256$ kV (line to line)

$$\delta = \tan^{-1}(0.22) = 12.4°$$

From $V_{LL} = 0.612 \times m_a \times V_{DC}$:

$$256 = 0.612 \times m_a \times 500$$
$$m_a = 0.84$$

Even though a controller based on Equations (11.10) and (11.11) could be used, in practice, controllers are developed by transforming the voltage and current into a d-q rotating reference frame as shown in Figure 11.25. The d-q frame is locked, using a Phase Locked Loop (PLL), onto the phasor of the phase-a in the AC system. The d-axis is chosen as the direction of phase-a voltage, thus $V_{gq} = 0$. The controller has two control loops, where the outer control loop controls the DC voltage and reactive power whereas the inner current control loops regulate the d and q components of the currents. On a HVDC system, one terminal is assigned the control of the DC voltage whereas the other terminal controls the real power flow (outer control loops control active and reactive power).

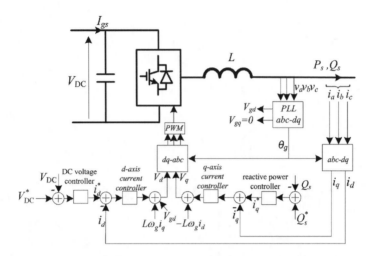

Figure 11.25   d-q controller for VSC

11.3.3   Multi-terminal HVDC

A Multi-terminal HVDC consists of a number of converters which are connected to a common HVDC circuit. They have been discussed since 1963 [12] when the first parallel multi-terminal HVDC system was proposed. Two multi-terminal HVDC links based on CSC-LCC were commissioned. The first one was the Sardinia–Corsica–Italy link which was converted from mercury arc rectifiers to thyristor-based CSC-LCCs in the 1990s. The second one was the Quebec–New England link which has a power rating of 2000 MW and was fully commissioned in 1992. More recently, the flexibility offered by VSC and the possibility of tapping HVDC links for wind farm connection have renewed interest in multi-terminal systems.

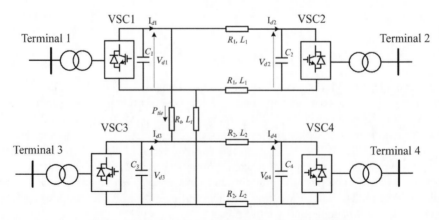

Figure 11.26   Multi-terminal HVDC with four terminals

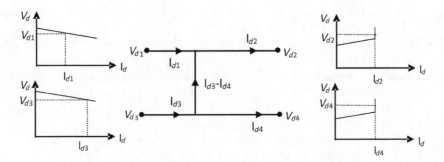

Figure 11.27   Droop controller for the four-terminal HVDC circuit

VSC-SCC is more suitable than CSC-LCC for multi-terminal HVDC as VSC can control the direction of the power flow without changing the polarity of the voltage. VSC-based multi-terminal HVDC is likely to be the key transmission technology for the proposed European Supergrid [13, 14].

A four-terminal HVDC is shown in Figure 11.26. The usual principle of operation is that one converter controls the DC voltage while the others regulate their power transfer using a controller shown in Figure 11.25. However, in order to set the power reference, communication between the converters may be required. Another method is to use a voltage droop control on all the terminals [15] as shown in Figure 11.27. Figure 11.27 shows the operating point of each VSC when power is delivered from the left-hand side to the right-hand side. The power sharing in the main HVDC lines is achieved by droop control. For example, if incoming power from VSC1 is increased, the controller in turn decreases V_{d1}. This reduces the power through the tie line as it is given by $(V_{d1} - V_{d3})/R_t$.

References

[1] *Our Electricity Transmission Network: A Vision for 2020*, ENSG report, March 2009, http://webarchive.nationalarchives.gov.uk/20100919181607/http:/www.ensg.gov.uk/assets/ensg_transmission_pwg_full_report_final_issue_1.pdf (accessed on 4 August 2011).

[2] *Oceans of Opportunity*, September 2009, http://ewea.org/fileadmin/ewea_documents/documents/publications/reports/Offshore_Report_2009.pdf (accessed on 4 August 2011).

[3] Gyugyi, L. (2000) Application characteristics of converter-based FACTS controllers. *International Conference on Power System Technology*, **1**, 391–396.

[4] Gyugyi, L. (1988) Power electronics in electric utilities: static VAR compensators. *Proceedings of the IEEE*, **76**(4), 483–494.

[5] Viawan, F.A., Sannino, A., Romero, I. and Maibach, P. (2008) *STATCOM for wind farms fault ride through improvements and grid code compliance*. 7th International conference on large scale integration of wind power and on transmission networks for offshore wind farms, 26–27 May, 2008, Madrid, Spain.

[6] Hingorani, N.G. and Gyugyi, L. (1999) *Understanding Facts: Concepts and Technology of Flexible AC Transmission Systems*, IEEE Press. London.

[7] Anderson, P.M., Agrawal, B.L. and Van Ness, J.E. (1989) *Subsynchronous Resonance in Power Systems*, IEEE Press, London.

[8] Tleis, N. (2007) *Power System Modelling and Fault Analysis: Theory and Practice*, Newnes, Oxford.

[9] Adamson, C. and Hingorani, N. (1960) *High Voltage Direct Current Power Transmission*, Garraway, London.

[10] Arrilaga, J., Liu, Y.H. and Watson, N.R. (2007) *Flexible Power Transmission: The HVDC Options*, John Wiley & Sons, Ltd, Chichester.

[11] Baker, C. *et al.* (2010) *HVDC: Connecting to the Future*, Alstom Grid.

[12] Lamm, U., Uhlmann, E. and Danfors, P. (1963) Some aspects of tapping HVDC transmission systems. *Direct Current*, **8**(5), 124–129.

[13] Asplund, G. (2009) *HVDC Grids: Possibilities and Challenges*, CIGRE SC B4 Bergen Colloquium, Norway.

[14] Gordon, S. (2006) Supergrid to the rescue. *IET Power Engineer*, **20**, 30–33.

[15] Haileselassie, T., Uhlen, K. and Undeland, T. (2009) *Control of multi-terminal HVDC transmission for offshore wind energy*. Nordic Wind Power Conference, Bornholm, Denmark, September 11, 2009.

12

Energy Storage

12.1 Introduction

Large quantities of electrical energy can be stored using pumped hydro or underground compressed air facilities. Such schemes can have a power rating of up to 1–2 GW with an energy capacity of 10–20 GWh. Smaller quantities of energy can be stored in batteries, flywheels and Superconducting Magnetic Energy Storage (SMES) devices [1–3]. Fuel cells convert a continuous source of chemical energy into electricity but have a similar impact on the power network as some energy storage systems (for example, flow batteries).

Figure 12.1 (also refer to Plate 9) shows the power and energy outputs of some of the electricity storage schemes that have been implemented using different technologies [1, 4, 5]. Fuel cells of a few hundred kW to several MWs are now in operation as a continuous source of power and are not shown in Figure 12.1. (Also refer to Plate 9) [6].

Applications of energy storage in the power system can be considered as being divided into those whose prime function is to deliver short-term power (kW) or those primarily supplying energy (kWh) over a longer period. Power quality, voltage support and some frequency support services which require short-term power support use batteries, flywheels and SMES which have a high power to energy ratio. Support for renewable energy, electrical energy shifting and end user energy management requires a large amount of energy and a discharge duration of several minutes to hours. Pumped hydro, compressed air storage, thermal energy storage and (flow) batteries are suitable candidate technologies. The initial market opportunities of energy storage will be developed in specific locations, but multiple income streams are likely to be necessary to cover the high cost of the storage equipment and account for the efficiency loss during charging and discharging.

Some of the applications of energy storage include:

1. *Power quality*: With the widespread use of sensitive electrical equipment, power quality is becoming an increasingly important phenomenon. Battery energy storage is used in Un-interruptible Power Supplies (UPSs) to mitigate short-term loss of power and power

Smart Grid: Technology and Applications, First Edition.
Janaka Ekanayake, Kithsiri Liyanage, Jianzhong Wu, Akihiko Yokoyama and Nick Jenkins.
© 2012 John Wiley & Sons, Ltd. Published 2012 by John Wiley & Sons, Ltd.

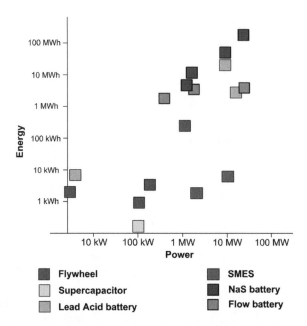

Figure 12.1   Implemented energy storage schemes based on technologies

fluctuations. They are commercially available and cost-effective now. Energy storage can also be used to mitigate voltage fluctuations and improve some other power quality issues such as harmonics. The alternative to energy storage for power quality applications is to make the control systems of the sensitive equipment more robust.

2. *Service provision to renewable generation*: Intermittent supply and lack of controllability are inherent characteristics of renewable energy generation. This is a challenge for the secure operation of the power system. Energy storage could support both the power system and renewable energy sources by smoothing their output, matching contract positions (or enabling scheduled dispatch) and shifting the generated energy in time. In a competitive energy market, renewable energy sources have the disadvantage of it being difficult to precisely forecast their output. Energy storage could reduce the error in the output forecasts of renewable generators by supplying the energy deficit or absorbing the excess.

The output of a renewable energy sources has no direct correlation to the electrical demand. Energy storage can store the energy when the resource is greater than the demand and supply the load when demand is greater than the supply. This is particularly useful when low demand conditions coincide with high wind power conditions.

Some power utilities are demonstrating the use of battery energy storage to smooth the output power of wind farms. For example, a NaS battery of 245 MWh, 34 MW has been installed at the 51 MW Rokkasho wind farm in Japan (shown in Figure 12.2, also refer to Plate 10). Figure 12.3 (also refer to Plate 11) shows how the battery energy storage is used to smooth the wind power output to meet a scheduled demand.

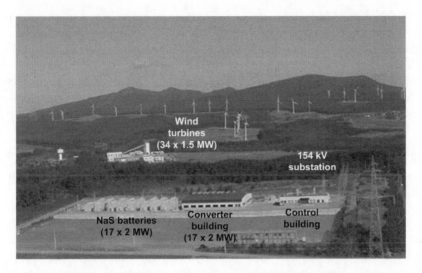

Figure 12.2   Rokkasho wind farm, Japan. *Source:* Courtesy of Japan Wind Development Co. Ltd.

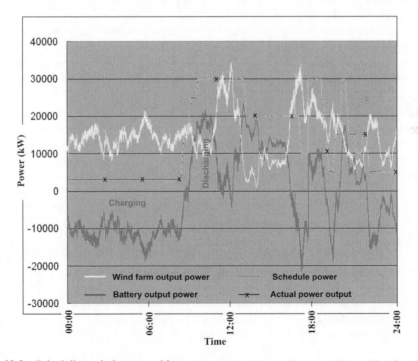

Figure 12.3   Scheduling wind power with energy storage. *Source:* Courtesy of Japan Wind Development Co. Ltd. Note that actual power output and scheduled power output coincide

The alternatives to energy storage to compensate for the variable output of renewable generation are fast response gas turbine generators, or interconnection of the renewable generation over a wide geographic area to smooth the output and Demand-Side Integration (Chapter 5).

3. *Electrical energy time shifting*: This involves storing energy during periods when demand is low or at times when the price is low, and discharging the energy when demand is high or at times when the price is high. This enables efficient utilisation of energy. It also supports distribution networks by relieving congestion during peak demand periods, providing the energy storage unit is positioned correctly within the distribution system.

4. *End use energy management*: This is energy management at the customers' premises. Energy storage could provide benefits to end users who are on time-of-use tariff through electrical energy time shifting or who have micro-generation.

   The main alternative to energy storage for energy time shifting and end use energy management is Demand-Side Integration.

5. *Voltage support*: Under normal operating conditions, system voltage is maintained within an upper and a lower limit. This is generally achieved by transformer tap changers and/or reactive power flows. In distribution circuits both active power and reactive power need to be used for voltage control due to the low X/R ratio of the network. Distributed energy storage may be attractive as it can provide both active and reactive power and control voltage while reducing the reactive power flows in the network.

   When connecting distributed generation, the minimum and maximum voltage limits of the distribution networks are a common limitation. If it is possible to use a form of active voltage control on the distribution network, then it may be possible to increase the maximum amount of distribution generation that can be connected. A STATCOM with energy storage on its DC side may be employed for voltage control, as discussed in [7].

6. *Reserve*: This is an ancillary service that is maintained to ensure system stability under unexpected connection/disconnection of load/generation. A fraction of the reserve is spinning and supplied by part-loaded large generators. These part-loaded generators operate at reduced efficiency due to the thermal losses incurred. Energy storage, if used to replace spinning reserve, does not discharge on a regular basis but is available to discharge if needed.

   As the amount of wind generation in an electricity network increases, and the uncertainties in wind output start to become evident, some extra balancing services will be required. It is anticipated that energy storage, such as from electrical vehicles, will play an important role in this.

   In addition to spinning reserve, the balancing task can be supported by a so-called standing reserve, which is supplied by higher fuel cost plant, such as Open Cycle Gas Turbines (OCGT) and energy storage. The advantage of storage over OCGTs lies in its ability to exploit (store) surpluses in generation during periods of high wind and low demand, and subsequently make a part of this energy available, and hence reduce costs. Further, storage can store and release energy while the OCGT plant can only provide energy to the power system.

7. *Load following*: This is an ancillary service purchased by some utilities to follow frequently changing power demand. As storage devices can be operated at partial output levels with high efficiencies and their response is quick, energy storage could be an ideal candidate for this service.

8. *Capacity of distribution circuits*: The increase in distributed generation and with electricity becoming the main energy vector will cause distribution lines to be overloaded, thus demanding distribution circuit upgrades. Energy storage could be used to relieve the congestion of distribution circuits and defer circuit reinforcement.

12.2   Energy storage technologies

12.2.1   Batteries

Batteries store energy in chemical form during charging and discharge electrical energy when connected to a load. In its simplest form a battery consists of two electrodes,[1] a positive and a negative placed in an electrolyte. The electrodes exchange ions with the electrolyte and electrons with the external circuit.

Lead acid and Sodium Sulfur (NaS) batteries are used at present for large utility applications in comparable numbers. Lithium Ion (Li-ion), Nickel Cadmium (NiCd) and Nickel metal hydrides (NiMH) are also thought to be promising future options.

Lead acid batteries have been used for many years in utility applications, providing excitation for synchronous machines and acting as back-up auxiliary power supplies. They are cheap but need significant maintenance. Their lifetime is comparatively short particularly if discharged deeply.

NaS batteries operate at 300–400 °C and have a large energy capacity per unit volume and weight. They are used for electrical energy time shifting (for example, Citizens Substation, USA, 2 MW (Figure 12.4), 12 MWh), wind farm support (for example, East Busco Substation, USA, 1MW, 6 MWh) and to smooth the output of PV generators.

Figure 12.5 is a simplified drawing of a NaS battery. The positive electrode is molten sulfur and the negative electrode is molten sodium. The electrolyte, which is a sodium beta-alumina ceramic (four layers of oxygen atoms and aluminium in the same atomic arrangement), allows ion exchange to take place. More details about the operation of a NaS battery are given in [3].

Li-ion is gaining its place as a distributed energy storage system, in particular with the developing use of Li-ion batteries in electric vehicles. These have a graphite negative electrode and lithium cobalt oxide, lithium iron phosphate or lithium manganese oxide positive electrode. The electrolytes generally use lithium salt in an organic solvent.

NiCd batteries were extensively used for power tools, mobile phones and laptops. However, this material has effectively been displaced from these markets by other batteries over the past decade. The robust nature of the technology, combined with its energy density, gives it advantages over lead acid and led to NiCd being chosen for the flagship battery energy storage project in Golden Valley in Alaska.[2]

---

[1] The positive electrode is the terminal where electrons flow into the battery (or current flows out) through the external circuit. The negative electrode produces electrons into the external circuit. The positive electrode is also called the cathode and the negative electrode is called the anode.

[2] Golden Valley Electric Association: Access at: http://www.gvea.com/about/bess/ (viewed on 4 August 2011).

Figure 12.4 Rock County Substation, Beaver Creek, Minnesota: 1 MW, 6 MWh NaS battery-based storage system installed at the point of interconnection of the Minwind 12 MW wind farm. *Source:* Courtesy of S & C Electric Europe Ltd.

12.2.2   Flow battery

A flow battery uses two electrolytes, often different kinds of the same chemical compound. Both the positive and negative electrolytes are stored separately and are pumped through a cell. Inside the cell, the two electrolytes are kept separate. The electrochemical reaction takes place by transferring ions across a membrane as shown in Figure 12.6. The electrodes do not take part in the chemical reaction and thus do not deteriorate from repeated cycling (one of the problems in other types of batteries that leads to loss of performance over time).

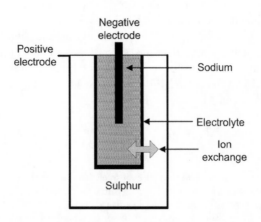

Figure 12.5   A NaS battery

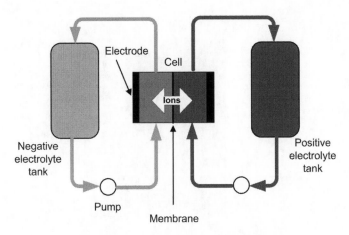

Figure 12.6   Construction of a flow battery

The amount of energy stored in a flow battery depends on the volume of the electrolyte in the tanks whereas the power output depends on the speed of ion transfer across the membrane.

Flow batteries using Zinc Bromide (ZBB) and Vanadium Redox (VRB) are available. A ZBB consists of a zinc negative electrode and a bromine positive electrode separated by a micro-porous membrane. An aqueous solution of zinc bromide (ZnBr) is circulated through the two compartments of the cell from two separate reservoirs as shown in Figure 12.6. On discharge, the zinc is oxidised, giving zinc ions, and the bromine is reduced to bromide ions. During charging, zinc is electroplated on the negative electrode and bromine is evolved at the positive electrode; this is stored as a chemically complex organic phase at the bottom of the positive electrolyte tank. A third pump is used for recirculation of the organic phase during the discharge cycle.

The reactions that occur at the two electrodes during charge and discharge are:

$$\text{Positive electrode: } Br_2(aq) + 2e^- \underset{\text{Charge}}{\overset{\text{Discharge}}{\rightleftharpoons}} 2Br^-$$

$$\text{Negative electrode: } Zn \underset{\text{Charge}}{\overset{\text{Discharge}}{\rightleftharpoons}} Zn^{2+} + 2e^-$$

In VRB, V^2/V^3 and V^4/V^5 redox couples in sulphuric acid are stored in electrolytic tanks with an ion exchange membrane. The reactions that occur in the battery during charging and discharging are:

$$\text{Positive electrode: } \quad V^{5+} + e^- \underset{\text{Charge}}{\overset{\text{Discharge}}{\rightleftharpoons}} V^{4+}$$

$$\text{Negative electrode: } V^{2+} \underset{\text{Charge}}{\overset{\text{Discharge}}{\rightleftharpoons}} V^{3+} + e^-$$

There have been a number of demonstration projects with flow batteries of capacities ranging from a few kW to several MW with storage up to 10 hours at full output [5].

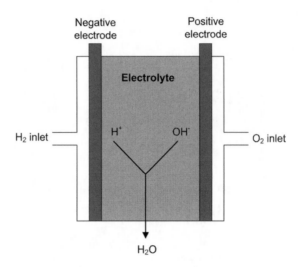

Figure 12.7 Simplified diagram of a fuel cell

12.2.3 Fuel cell and hydrogen electrolyser

Most fuel cells use H_2 and O_2 as their main fuel (Figure 12.7). At the negative electrode, hydrogen is oxidised to form H^+ and electrons. The electrons flow through the external electrical circuit whereas the hydrogen ions move towards the positive electrode through the electrolyte. The positive electrode is made from a porous material coated with a catalyst. At that electrode, the hydrogen ions combine with oxygen to produce water.

Many different types of fuel cells have been developed and their characteristics are listed in Table 12.1.

Table 12.1 Different types of fuel cells [8]

Type	Electrolyte	Operating temperature	Maximum output power reported
Polymer Electrolyte Membrane (PEM)	Organic polymer	80 °C	250 kW
Alkaline Fuel Cells (AFC)	Potassium hydroxide	150–200 °C	12 kW
Phosphoric Acid Fuel Cell (PAFC)	Phosphoric acid	150–200 °C	250 kW
Molten Carbonate Fuel Cells (MCFC)	Potassium, sodium or lithium carbonate	650 °C	2 MW
Solid Oxide Fuel Cells (SOFC)	Ceramic materials	1000 °C	100 kW

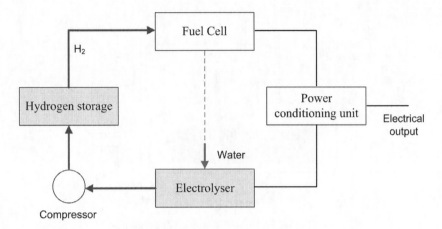

Figure 12.8   Hydrogen energy storage system

Hydrogen-based energy storage systems consist of an electrolyser, hydrogen storage and a fuel cell, as shown in Figure 12.8. The electrolyser uses electrical energy to produce H_2 from water. One of the potential applications of this device is to store H_2 when there is excess wind energy generation and then use the stored H_2 to support the power system during peak demand periods.

Regenerative fuel cells which consume electricity and act as an electrolyser (to produce H_2) have also been developed.

12.2.4   Flywheels

Flywheels store kinetic energy in a rotating mass and release it by slowing the rotation when electrical energy is required. Their application to date has mainly been for power quality and to provide energy for UPS. The majority of installations are on customers' premises with a very few applications in demonstration micro-grids.

Consider the simple rotating disc shown in Figure 12.9. The stored energy is proportional to the square of the rotor speed (ω) and the moment of inertia of the rotating mass (J):

$$\text{Stored energy} = \frac{1}{2}J\omega^2 \qquad (12.1)$$

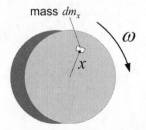

Figure 12.9   A rotating disc

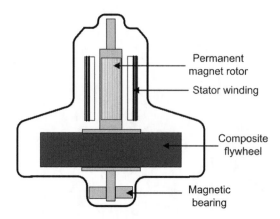

Figure 12.10   Cross-section of a flywheel

The moment of inertia is given by [3]:

$$J = \int x^2 \times dm_x \tag{12.2}$$

where x is the distance to an element of mass dm_x from the axis of rotation.

From Equation (12.1), it is clear that to store energy, two options can be used: a low speed flywheel which uses a heavy steel rotor with high inertia or a high speed flywheel which uses lighter composite materials.

As given in Equation (12.2), the location of the mass is important. In some designs the mass is formed as a rim, placed away from the axis of rotation, to maximise the moment of inertia.

High speed flywheels, which rotate at up to 50,000 rpm, operate in a vacuum with magnetic bearings in order to reduce friction losses. A cross-section of a flywheel is shown in Figure 12.10.

Example 12.1

A flywheel has a concentrated rim of carbon fibre. The inner radius of the rim (r_1) is 10 cm and the outer radius (r_2) is 23 cm. The thickness of the rim (l) is 20 cm. If carbon fibre has a density of 1500 kg/m^3, what is the moment of inertia of the flywheel? Hence calculate at what speed the flywheel should be rotated to obtain short-term energy storage of 1 kWh.

Answer

Figure 12.11 shows a flywheel.

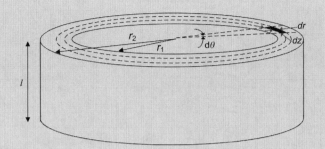

Figure 12.11   A flywheel

The mass of the infinitesimal section shows

$$dm_x = \rho l \times dz \times dr$$

where ρ is the density of the ring.
   Since

$$dz = r \times d\theta : \quad dm_x = \rho l \times r d\theta \times dr$$

From Equation (12.2), the moment of inertia of the ring is given by

$$J = \int_{r=r_1}^{r=r_2} \int_{\theta=0}^{2\pi} r^2 \times \rho l \times r d\theta \times dr$$

$$= \int_{r=r_1}^{r=r_2} 2\pi \rho l \times r^3 \times dr$$

$$= \pi \rho l \left[\frac{r_2^4 - r_1^4}{2} \right]$$

Substituting values:

$$J = \pi \times 1500 \times 0.2 \left[\frac{0.23^4 - 0.1^4}{2} \right] = 1.27 \text{ kgm}^2$$

From Equation (12.1):

$$\text{Stored energy} = \frac{1}{2} \times 1.27 \times \omega^2 = 1 \times 10^3 \text{ W} \times 3600 \text{ s}$$

Therefore, the speed of rotation should be 2381 rad/sec (22,737 rpm).

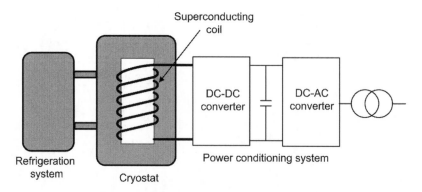

Figure 12.12   Component of a superconducting energy storage

12.2.5   Superconducting magnetic energy storage systems

In a SMES system, a magnetic field is created by direct current passing through a superconducting coil (Figure 12.12). In a superconducting coil, resistive losses are negligible and so
the energy stored in the magnetic field (equal to $LI^2/2$ where L is the inductance of the coil
and I is the current passing through the coil) does not reduce with time. However, in order to
maintain the superconductivity of the SMES coil, a cryostat which can keep the temperature
of the coil below the superconductor temperature limit is required. The optimum operating
temperature of high temperature superconductors, that are favoured for energy storage applications, is around 50–70 K. Further, as the magnetic field produced by a SMES is large, a
strong supporting structure is needed to contain the electromagnetic forces. The stored energy
in the SMES is retrieved when required by a power conditioning system that is connected to
the AC network as shown in Figure 12.12.

12.2.6   Supercapacitors

In a capacitor, energy is stored in an electrostatic field. The quantity of energy stored, E, is
given by:

$$E = \frac{1}{2}CV^2$$

where C is the capacitance and V is the voltage across the capacitor.
In a parallel plate capacitor:

$$C = \frac{\varepsilon_r \varepsilon_o A}{d}$$

where ε_r is the relative permittivity of the dielectric material, ε_o is the permittivity of free
space, A is the surface area of the plates and d is the distance between two plates.

Supercapacitors have a double layer structure that uses a porous electrolyte such as Polyethy-
lene Terephathalate (PET). The double layer structure increases energy storage capability
significantly due to a large increase in surface area, thus C. Further, the higher relative

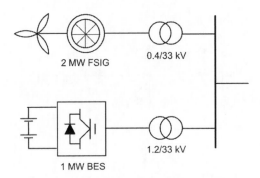

Figure 12.13 System used for the case study

permittivity of PET (about 3.5 at 20 °C) compared to the electrolytes used in thin film capacitors also contributes to the increase in stored energy. Supercapacitors are constructed with a carbonised porous material as one electrode and a liquid chemical conductor as the other electrode.

12.3   Case study 1: Energy storage for wind power

Figure 12.13 shows a 2 MW fixed speed induction generator (FSIG) wind turbine connected to a Battery Energy Storage (BES).

The BES system uses a multi-modular converter shown in Figure 12.14a. The switches in each bridge are switched to obtain a step-wise output as shown Figure 12.14b.

During the positive half cycle, all the upper switches on the first arm of the bridges (S_1 to S_6) were on; whereas during the negative half cycle, all the lower switches on the first arm ($\bar{S}_1$ to $\bar{S}_6$) were on.

Within a half-cycle, staircase modulation was achieved by switching the switches of the second arm. In order to obtain a step of V_{dc}, one lower switch of the second arm was turned on. For example, when S_{11} was on and all the other lower switches of the second arm (that is, S_{21}, S_{31}, S_{41}, S_{51} and S_{61}) were off, the output, V_o, was equal to V_{dc}. To obtain a step of $2V_{dc}$, two lower switches of the second arm were turned on. Finally, to obtain $6V_{dc}$, all the lower switches of the second arm of the bridges were turned on.

Table 12.2 shows one possible combination of switching status during the positive and negative half cycles. The switches which were turned on were cyclically varied so as to extract equal amounts of energy from each battery. That is, during the first cycle, Bridge 1 was used to obtain V_{dc}; whereas during the second cycle, Bridge 2 was used to obtain V_{dc}. Turn-on times ($\beta 1$, $\beta 2$, and so on) were obtained by an optimisation routine where the error between the waveform shown in Figure 12.14(b) and a sinusoidal waveform was minimised.

Using the BES shown in Figure 12.14a, the system shown in Figure 12.13 was simulated. As shown in Figure 12.15a, a varying wind speed was applied to the fixed speed wind turbine. The simulations started at 6 a.m. and finished at 6 a.m. on the next day. From 6 a.m. to 10 p.m. the BES was controlled to maintain the total power supplied to the grid at 2 MW as shown in Figure 12.15d. From 10 p.m. to 6 a.m. the following day, the load was set to zero and the wind power was used to charge the BES. Figure 12.15e shows the State Of Charge (SOC) of

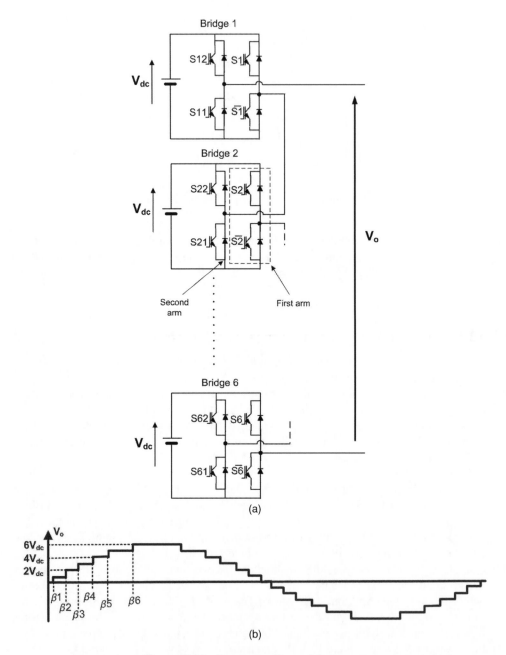

(a)

(b)

Figure 12.14 Output waveform of the inverter

Table 12.2 One possible switching combination

	During the positive half cycle								
V_o	S_1 to S_6	$\bar{S}_1$ to $\bar{S}_6$	S_{11}	S_{21}	S_{31}	S_{41}	S_{51}	S_{61}	S_{12} to S_{62}
0	0	0	0	0	0	0	0	0	0
$V_{dc}/6$	1	0	1	0	0	0	0	0	0
$2V_{dc}/6$	1	0	1	1	0	0	0	0	0
$3V_{dc}/6$	1	0	1	1	1	0	0	0	0
$4V_{dc}/6$	1	0	1	1	1	1	0	0	0
$5V_{dc}/6$	1	0	1	1	1	1	1	0	0
$6V_{dc}/6$	1	0	1	1	1	1	1	1	0

	During the negative half cycle								
	S_1 to S_6	$\bar{S}_1$ to $\bar{S}_6$	S_{12}	S_{22}	S_{32}	S_{42}	S_{52}	S_{62}	S_{11} to S_{61}
$-V_{dc}/6$	0	1	1	0	0	0	0	0	0
$-2V_{dc}/6$	0	1	1	1	0	0	0	0	0
$-3V_{dc}/6$	0	1	1	1	1	0	0	0	0
$-4V_{dc}/6$	0	1	1	1	1	1	0	0	0
$-5V_{dc}/6$	0	1	1	1	1	1	1	0	0
$-6V_{dc}/6$	0	1	1	1	1	1	1	1	0

each battery bank of the six-level converter. As can be seen, the SOC reduced from 6 a.m. to 10 p.m. and then increased. The SOC in each bridge (SOC1 is the state of charge of the battery in Bridge 1, and so on) was balanced by the cyclic modulation technique.

12.4 Case study 2: Agent-based control of electrical vehicle battery charging

Many countries are promoting Electric Vehicles (EV) as a means of decarbonising their transport sectors. From the power system point of view, EVs can be viewed not only as loads but also as distributed energy storage devices. It is anticipated that EVs will communicate with the Smart Grid to provide electrical energy demand-shifting services such as reducing their charging rate or delivering electricity to the grid.

In this case study, two charging regimes, uncontrolled EV charging (where all the consumers charge their EVs just after returning home) and controlled EV charging, were investigated. A Multi Agent System (MAS) was used for controlled EV charging. More information about MAS can be found in [9, 10].

A residential area having Battery Electric Vehicles (BEV) and Plug-in Hybrid Electric Vehicles (PHEV)[3] was considered. It was assumed that there are 32 BEVs and 96 PHEVs. The mean time the EVs were connected for charging was assumed to be 18.00 hrs. The parameters of the EVs are given in Table 12.3.

---

[3] Both types are referred to as EVs.

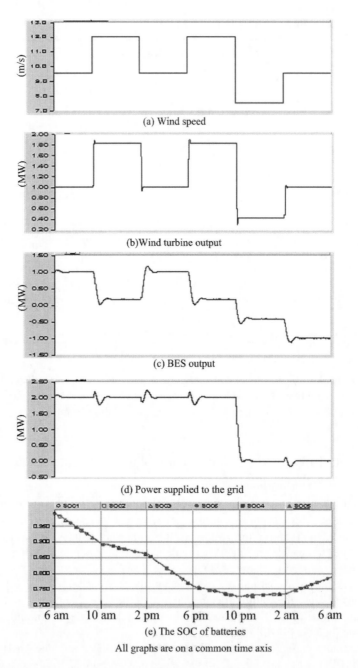

(a) Wind speed

(b) Wind turbine output

(c) BES output

(d) Power supplied to the grid

(e) The SOC of batteries

All graphs are on a common time axis

Figure 12.15   BES output and SOC level of batteries for different wind speeds

Table 12.3 Assumed EV parameters

BEV battery capacity (kWh)	28
PHEV battery capacity (kWh)	7.2
EV charge rating (kW)	2.99
EV battery efficiency (%)	85
EV charger efficiency (%)	87
Average EV energy requirement (kWh)	6.5
Average BEV initial State of Charge (%)	40
Average PHEV initial State of Charge (%)	10

An agent architecture was used (Figure 12.16).

1. The Substation Agent (**SA**) is located at the secondary substation at the MV/LV level. It is responsible for managing the battery charging/discharging of the electric vehicles within a LV area.
2. The Vehicle Agent (**VA**) is located at the vehicle and represents an EV owner or an EV Aggregator.[4] It is responsible for managing the charging/discharging of an individual EV or fleet of EVs.

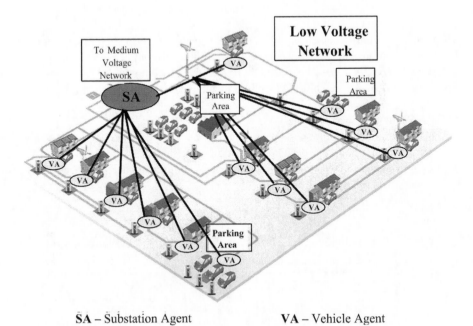

SA – Substation Agent   VA – Vehicle Agent

Figure 12.16 Multi Agent-based Electrical Vehicle charging and discharging architecture

---

[4] An Aggregator for EVs is a commercial entity operating between the system operator and the EVs. The Aggregator represents a large number of EVs so that the system operator sees the collection of EVs as a large generator or load.

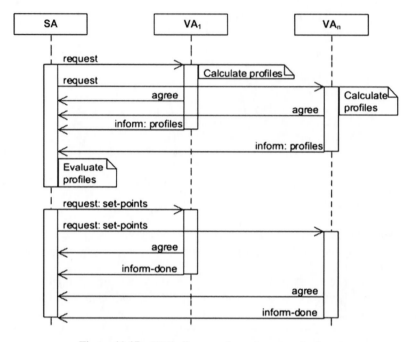

Figure 12.17   UML diagram of agent communication

The case study considers the case when the **SA** detects an overload on the transformer of the MV/LV substation. The policy examined is to avoid the violation of the loading limit which is assumed to be 700 kVA. The procedure followed is based on the FIPA Request Interaction Protocol and is shown in Unified Modelling Language (UML) notation in Figure 12.17.

- The **SA** initiates requests to the **VA**s.
- The **VA**s respond to the requests providing a number of possible charging schedules based on the EV owner preferences and the EV equipment characteristics.
- The **SA** evaluates the responses based on the loading limit and decides the charging profiles that need to be followed.

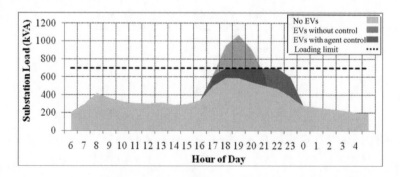

Figure 12.18   Substation Load Demand

Figure 12.18 shows the load demand at the MV/LV transformer with the EV charging. As can be seen, uncontrolled EV charging resulted in doubling the demand. Using the MAS (EVs with agent control), EV demand could be maintained within the transformer rating.

References

[1] Sarrias, R., Fernández, L.M., García, C.A. and Jurado, F. (2010) Energy storage systems for wind power application. International Conference on Renewable Energies and Power Quality (ICREPQ'10), 23–25 March, 2010, Granada, Spain.

[2] *An Assessment of Battery and Hydrogen Energy Storage Systems Integrated with Wind Energy Resources in California*, Public Interest Energy Research Program, California Energy Commission, September 2005.

[3] Ter-Gazarian, A. (1994) *Energy Storage for Power Systems*, IEE Energy Series **6**.

[4] Roberts, B. (2009) Capturing grid power: performance, purpose, and promise of different storage technologies. *IEEE Power and Energy Magazine*, **7**(4), 32–41.

[5] Guerrero, M.A., Romero, E., Barrero, F. et al. *Supercapacitors: Alternative Energy Storage Systems*, http://peandes.unex.es/archives%5CP126.pdf (accessed on 4 August 2011).

[6] Curtin, S., Gangi, J. and Delmont, E. (2011) *State of the States: Fuel Cells in America*, June 2011, http://www.fuelcells.org/StateoftheStates.pdf (accessed on 4 August 2011).

[7] Arulampalam, A., Ekanayake, J.B. and Jenkins, N. (2003) Application study of a STAT-COM with energy storage. *IEE Proceedings: Generation, Transmission and Distribution*, **150**(3), 373–384.

[8] *Fuel Cell Handbook* (5th edition), EG&G Services Parsons, Inc., 2000, http://www.fuelcells.org/info/library/fchandbook.pdf (accessed on 4 August 2011).

[9] Wooldridge, M. and Jennings, N.R. (1995) Intelligent agents: theory and practice. *The Knowledge Engineering Review*, **10**(2), 115–152.

[10] Bellifemine, F., Caire, G. and Greenwood, D. (2007) *Developing Multi-Agent Systems with JADE*, John Wiley & Sons, Ltd, Chichester.

Index

Accuracy
current transformers, 116
voltage transformers, 121
Active distribution network, 7
Active filter, 224–5
Actuator
ring main unit, 127–8
Addressing
classful, 41–2
classless, 42
Adjustable speed drivers, 109
Agents, 133, 275–6
AM/FM/GIS, 168
Analogue to digital conversion (ADC)
accuracy, 89
resolution, 90–1
sigma delta, 92–3
successive approximation, 92
Ancillary services, 107
ANSI C12.22, 63
Apparent power, 94
Asset
ageing, 2
management, 7
Asynchronous generator (*see* Generators)
Attenuation, 26
Authentication, 76–7
Automatic meter reading (AMR), 85
Automatic meter management (AMM), 85
Automatic voltage controller (AVC), 136
Automatic voltage regulators (AVR), 9

Bandwidth, 26, 32–4, 47
Base values (*see* Per unit system)
Battery
flow, 264–6
lead acid, 263
lithium ion, 263
nickel cadmium, 263
sodium sulphur, 263–4
Bay controller, 20–1, 115, 124–5
Bit rate, 26–9, 47–8, 54, 59–60
Bluetooth, 53–4

Cables
coaxial, 29
optical fibre, 29–32
unshielded twisted pair, 28
Capacitor voltage transformer (*see* voltage transformers)
Cipher
substitution, 71
transposition, 71–2
Circuit breaker, 127–9, 132–3
Circuit switching, 24
Climate change, 1
CO_2 emission, 83, 111, 206
Common Information Model (CIM), 142
Communication
channels, 19–22, 25–6
mobile, 59–60
protocol, 35–41, 98–9

Smart Grid: Technology and Applications, First Edition.
Janaka Ekanayake, Kithsiri Liyanage, Jianzhong Wu, Akihiko Yokoyama and Nick Jenkins.
© 2012 John Wiley & Sons, Ltd. Published 2012 by John Wiley & Sons, Ltd.

Printed and bound by CPI Group (UK) Ltd, Croydon, CR0 4YY